**B F695d**
**Dierikx, M. L. J.**
 **Fokker**

# FOKKER

# SMITHSONIAN HISTORY OF AVIATION SERIES
## Von Hardesty, Series Editor

On December 17, 1903, human flight became a reality when Orville Wright piloted the *Wright Flyer* across a 120-foot course above the sands at Kitty Hawk, North Carolina. That awe-inspiring twelve seconds of powered flight inaugurated a new technology and a new era. The airplane quickly evolved as a means of transportation and a weapon of war. Flying faster, farther, and higher, airplanes soon encircled the globe, dramatically altering human perceptions of time and space. The dream of flight appeared to be without bounds. Having conquered the skies, the heirs to the Wrights eventually orbited the Earth and landed on the Moon.

Aerospace history is punctuated with many triumphs, acts of heroism, and technological achievements. But that same history also showcases technological failures and the devastating impact of aviation technology in modern warfare. As adapted to modern life, the airplane—as with many other important technological breakthroughs—mirrors the darker impulses as well as the genius of its creators. For millions, however, commercial aviation provides safe, reliable, and inexpensive travel for business and leisure.

This book series chronicles the development of aerospace technology in all its manifestations and subtlety. International in scope, this scholarly series includes original monographs, biographies, reprints of out-of-print classics, translations, and reference materials. Both civil and military themes are included, along with systematic studies of the cultural impact of the airplane. Together, these diverse titles contribute to our overall understanding of aeronautical technology and its evolution.

# FOKKER

## A TRANSATLANTIC BIOGRAPHY

### MARC DIERIKX

SMITHSONIAN INSTITUTION PRESS

WASHINGTON AND LONDON

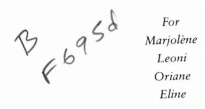

*For*
*Marjolène*
*Leoni*
*Oriane*
*Eline*

Copy editor: Tom Ireland
Production editor: Jack Kirshbaum
Designer: Kathleen Sims

Library of Congress Cataloging-in-Publication Data

Dierikx, M. L. J.
    Fokker : a transatlantic biography / Marc Dierikx.
        p.        cm.
    Includes bibliographical references (p.     ) and index.
    ISBN 1-56098-735-9 (cloth)
    1. Fokker, Anthony H. G. (Anthony Herman Gerard), 1890-1939.
    2. Aeronautical engineers–Netherlands–Biography. I. Title.
TL540.F6D54 1997
629.13'0092–dc21
    [B]                                                      97-4296

British Library Cataloguing-in-Publication Data is available

Manufactured in the United States of America
04 03 02 01 00 99 98 97   5 4 3 2 1

⊗ The paper used in this publication meets the minimum requirements of the American
National Standard for Information Sciences—Permanence of Paper for Printed Library
Materials ANSI Z39.48-1984. Printed on recycled paper.

Jacket illustration: Anthony Fokker posing under the wing of one of KLM's five Fokker
F.XVIII airliners in 1932.

For permission to reproduce illustrations appearing in this book, please correspond
directly with the author. The Smithsonian Institution Press does not retain reproduction
rights for these illustrations individually.

# CONTENTS

# PREFACE

Among the names that stand out in the history of prewar international aeronautical development, that of "Flying Dutchman" Anthony H. G. Fokker looms large. In the twenty years following aviation's rapid coming of age after the outbreak of war in August 1914, airplanes bearing his name counted among the best that aircraft technology had to offer. Yet by the mid-1930s his company appeared to be in terminal decline. Fokker's rise and demise as a constructor were equally remarkable, exemplifying wider developments in the field. From his beginnings as an unimportant youngster building sports planes in Germany, luck and wartime developments pushed Fokker to the fore. By 1924 his manufacturing and business interests stretched across the Atlantic from Germany and Holland to the United States. Until the end of the decade, Fokker was easily the largest aircraft producer in the world, and Fokker planes dominated aeronautical construction in Europe and the U.S. alike. After that, his star began to fade with the growth of the indigenous American aviation industry. But though it might have looked from a distance as if Fokker's days had passed, such appearances were deceptive. On the contrary, Tony Fokker continued to play an important role in the history of aviation through sales activities for companies such as Douglas and Lockheed that were instrumental in bringing about the worldwide hegemony of American-built aircraft.

The present book looks at the background of these developments, connecting them with the extraordinary personality of a man who became a
legend in his own lifetime, and with the times in which he lived. Because
this is a biography of an aircraft constructor, airplanes take up a substantial part of the attention. In that sense, the present book addresses the
history of the emerging technology of aeronautical engineering. However,
biographies are about people, their actions, their life. The specialist
reader will find that some technical developments are only covered su-
perficially, and that a number of the aircraft types that the Fokker factories turned out are not mentioned, or mentioned only in passing. The
same reader will look in vain for appendixes with technical data on
the airplanes. These have, in the past, been covered extensively and are
readily available in such standard publications as Michael J. H. Taylor's
*Jane's Encyclopaedia of Aviation* (1989). Instead, the present book attempts to bring out the person of Anthony Fokker in the context of his
various endeavors. This was no easy task, since very few personal records
providing meaningful insight to Fokker's personality have survived. The
somewhat fragmented picture that emerges reflects both the author's prolonged grappling with what documents could be recovered and the rest-
less nature of Fokker, a man who was forever on the move. Nevertheless,
this book seeks to provide an answer to the questions of what sort of a
person Anthony Fokker was, why he became an aviator, constructor, and
industrialist, and how he achieved that status. This means that the book
is also a business history of sorts. As far as sources permit, it reconstructs
Fokker's entrepreneurial policy and explains the way his personality and
financial strategy affected the sales of the various types of Fokker aircraft.
This is not the first book to deal with Anthony Fokker. On the basis of his
own recollections, Fokker had his autobiography, a somewhat shallow
depiction of his life recounting glorified wartime memories, written down
for him by journalist Bruce Gould in 1929–30. It appeared as *Flying
Dutchman: The Life of Anthony Fokker* and was first published by Henry
Holt and Company in New York in April 1931. An instant success, the
book was translated in several other languages and reprinted numerous
times. Fokker himself edited a Dutch version, adding some significant
personal details to the story. *Flying Dutchman* set the tone for future
works on Fokker's life, such as Edmond Franquinet's epigonic *Een leven
voor de luchtvaart* (A life for aviation) of 1946, and Henri Hegener's
well-illustrated but unperceptive *Fokker: The Man and the Aircraft* of
1961. If these books revered Anthony Fokker for his achievements,

Alfred Weyl's posthumous *Fokker: The Creative Years* of 1965 set out to do just the opposite. An impenetrable tangle of detailed facts and persuasive fiction, Weyl's book was written with one goal: to "do justice" to Fokker's welder-become-constructor, Reinhold Platz, who, Weyl was convinced, had been pivotal in Fokker's ascent in aviation and had been maliciously denied due credit in *Flying Dutchman*. Curiously enough, subsequent books dealing with Fokker—Thijs Postma's *Fokker: Aircraft Builders to the World* (1980), Peter Alting's *Van Spin tot Fokker 100* (1988) and *Fokkers in Uniform* (1988), Peter van de Noort's *Fokkers "Roaring Twenties"* (1988), Alex Imrie's *The Fokker Triplane* (1992), and Jörg Kranzhof's *Flugzeuge die Geschichte machten: Fokker Dr.I* (1994)— have nearly exclusively focused on Fokker airplane technology and devoted little attention to the man they were named after. The present book aims to set the record straight.

The research for this book has taken a long time. The starting point lay somewhere in the autumn of 1984, when the secretary of the board of the present Fokker Aircraft BV in Amsterdam, at that time R. W. van Eck, mentioned in conversation that the old archives of the company had been deliberately destroyed in 1947 to present the new management with a clean slate in rebuilding the company after the initial purges of Nazi collaborators. Only little bits and pieces of archival material that surfaced from desk drawers in later years had escaped the fire drills of the Amsterdam fire brigade, he said. From that moment on, my curiosity was aroused. Having the good fortune to be able to do various research projects in the field of aviation history since then, I made notes and copies of every bit of information on Fokker, and on his company, that I came across in books, archives, and libraries. They are listed in the bibliography. Over the years, my notes accumulated to the stage where it became possible to write the present book.

It is customary to thank those who have been instrumental in helping the author along the long road towards completion of a book. Such public gratitude is especially appropriate in this case. With the virtual absence of a central archival collection of documents from which to start compiling the story, I owe very much to the willingness of numerous individuals to point me in the direction of possible sources of information. Had it not been for their help and, often, hospitality, this book would not have materialized.

First and foremost, I wish to express my deep gratitude to Peter M. Grosz of Princeton, New Jersey, who has encouraged me to persist in my quest from the very start. A specialist in German and Austro-Hungarian aviation up to 1919, he generously did what so few researchers can persuade themselves to do, allowing me to use his personal notes and files compiled over decades of study. He also kindly agreed to proofread the first chapters of the book, and his comments saved me from a number of pitfalls.

A lot of the research for this book was done in Holland, and I am indebted to a number of people there for advice and assistance. Over the course of my researches three consecutive company historians of Fokker Aircraft in Amsterdam have helped me trace various pieces of information in company records and offered me the benefit of their specialist knowledge. They allowed me to use their photocopier and supplied me with numerous cups of coffee and lunches. I thank them in the order in which we first met: Peter Alting, Leo de Roo, Paul Moreu. The latter was also invaluable in supplying the collection of photographs that appears in this book. Aviation journalist and illustrator Thijs Postma of Hoofddorp was kind enough to receive me several times in his home and allowed me to use the personal papers of the late Henri Hegener that are in his possession. Peter van de Noort of the Aviodome Air and Space Museum at Schiphol Airport facilitated access to the Aviodome collection of documents and artifacts pertaining to Anthony Fokker and was instrumental in retrieving various files that I would not have found otherwise. I owe much to the cooperation of KLM Royal Dutch Airlines. The now retired international vice president, Henri A. Wassenbergh; and former secretary of the board, Jan-Willem Storm van 's-Gravesande, granted me access to the well-preserved Board Papers, which provided vital information on the relation between Fokker and KLM—so important for the development of the Dutch Fokker company in the interwar period. KLM's archivist, Dick Jansen, did most of the footwork of gathering documents and generally lent a keen ear to the stories as they emerged. Ton Nijland of The Hague, Fokker's nephew, spent hours recollecting his uncle. I am also grateful to Wim Teuben of Voorburg for receiving me at his home and making available Fokker artifacts and documents from his remarkable "Teubarium" collection. Gert Blüm of Haarlem put me on the right track in my initial researches on Fokker's American activities and also shared research notes. Sierk Plantinga of the Algemeen Rijks Archief in The Hague contributed numerous research suggestions, as did Walter Salzman of Leiden. Over the years I have been in touch with a number of other people in Holland interested in the history of Fokker. Several professional moves have made me lose track of their addresses, but I wish to thank them all cordially.

Further afield in Europe, Stefan Ittner of the Deutsches Museum in Munich, Germany, also offered research suggestions. Helmuth Trischler, director of research at the Deutsches Museum, pointed me to Fokker material in the Junkers Archive. Jean Roeder, senior vice president of Airbus Industrie in Blagnac, France, was kind enough to share some of his findings on the history of German aviation and made photocopies available from his files. In the United States, Ron Davies, Dominick Pisano, Tom Crouch, and Von Hardesty of the Smithsonian Institution's National Air and Space Museum were instrumental in getting this book actually written by offering me a Guggenheim Fellowship and sharing some of their wealth of knowledge with me. William F. Trimble of Auburn University, Alabama, gave me lots of friendly encouragement and a great many useful hints, and dug up copies from obscure sources. W. David Lewis, James R. Hansen, and Dixie Dysart, also of Auburn University, offered kind advice as well. Richard Allen of Lewiston, Idaho, shared some of his research findings on the American Fokker companies with me. Hein Volker of Pierrefonds, Quebec, explained the intricacies of machine gun synchronization. And last, but certainly not least, I am deeply indebted to Tom Heitzman of Deansboro, New York. A chance encounter brought three photo albums from his "Stuffinder" collection to my attention that had belonged to Fokker's former cook, Christine Döppler. These albums also contained a selection of hitherto unknown notes and letters from Anthony Fokker, and from his mother, that added considerably to my understanding of the man. Tom allowed me to examine and use them for this biography. My good friend Charles Verkuylen compiled the index.

I also expressly wish to mention with gratitude the material and immaterial support of my various employers over the past decade: the Catholic University of Nijmegen, the Niels Stensen Foundation of Amsterdam, the Business History Unit of the London School of Economics, the Royal Netherlands Academy of Arts and Sciences in Amsterdam, Auburn University, and Utrecht University. Without the academic freedom to indulge in this biographical project outside my regular duties, the research for this book could not have been undertaken. In a time in which academic budgets have shrunk time and time again, such freedom is becoming ever more precious.

ROSMALEN, THE NETHERLANDS

# 1

## GROWING WINGS

### BIRTHMARKS

The weather was hot and humid in Blitar, on the tropical island Java in the Dutch East Indies, and had been so for weeks. The monsoon season was drawing to an end, when, on April 6, 1890, Anna Fokker-Diemont gave birth to her second child, Anthony (Tony). She was twenty-four years old then. Three years earlier, she had arrived on Java by ship from Holland with her newlywed husband, Herman Fokker. Herman, the seventh and youngest child of the wealthy merchant, broker, and tanner Anthony Herman Gerard Fokker (1809–74) and his wife, Maria van den Broecke, was the proud owner of a coffee plantation in the remote mountainous district of Kediri, southeastern Java.

Descending from a long line of influential merchants and shipowners, Herman Fokker had been born on January 26, 1851, in the quiet backwater town of Middelburg, provincial capital of Zeeland in the Netherlands.[1] His father was a prominent member of the city council and the chamber of commerce, and a commissioner of the Nederlandsche Handel Maatschappij (NHM), the Dutch colonial trade and banking institution. Indeed, the Fokkers had been part of the economic and social elite for several generations. Young Herman did not fit well in this close-knit circle. Unlike several of his elder brothers, who pursued careers in law and

1

academics, he grew up to become a professed rough-and-tough nonintellectual. In this attitude, Herman Fokker was not unique. Numerous upper-class families faced comparable problems with rebellious sons. The accepted thing was to encourage them to settle in the Dutch East Indies, hoping that they might fit into colonial society better than into the demanding social environment of Holland. Therefore, it was no surprise that Herman boarded one of the ships of the Rotterdamsche Lloyd destined for Java after the death of his father in 1874. With the help of his elder brother, Anthony (born in 1842), a senior official at the office of the NHM in Batavia, he used his part of his father's inheritance to buy a coffee plantation. Thus, at the age of twenty-three, he established himself as a coffee planter.

Coffee, tea, cane sugar, and indigo were highly profitable export products of nineteenth-century Java. In Europe, increasing wealth and a developing popular taste for coffee set the stage for rising demand. Growing and trading coffee beans was a lucrative business because it enjoyed extensive protection from the Dutch colonial authorities. Exports were channeled through the NHM, founded in 1824. Operating under royal patronage, the NHM acquired the monopoly on exporting coffee, cane sugar, and indigo from the Dutch East Indies, products that were auctioned off in Amsterdam at high profits. Coffee growing was anchored in the so-called Cultures System (Cultuurstelsel), under which government concessions grew these crops on leasehold property after all land had been appropriated by the colonial government in a decree of 1830. Indeed, so profitable was the trade that these colonial products provided one-fifth to one-third of the Dutch gross national income.[2]

Initially, the Cultures System allowed planters to be posted as supervisors on government-owned plantations. They were given highly remunerative contracts for delivery of their products to the NHM in Batavia. Labor costs were extremely low because the planters used native statute labor on a large scale. Small wonder, then, that these concessions were much sought after. Small wonder, also, under the nepotist colonial regime, that many of these profitable contracts went to family members of government officials and NHM personnel. For the European upper crust of society, the Cultures System provided a secure shortcut to riches. Yet, rising moral objections to the system led to its abandonment in 1870. Thereafter, plantations were sold off to private owners at the termination of the leasehold contracts.

Herman Fokker was one of the new generation of planters who came in as a result of this change in policy, entering the coffee business when plantations were beginning to be auctioned off to private owners. Though sources are lacking, he was probably among the first of the planters who owned their fields. After years of hard labor, in 1887 Herman returned to Holland to marry his cousin, Johanna (Anna) Hugenota Wouterina Wilhelmina Diemont, from Leeuwarden in the northern province of Friesland. Anna was fifteen years younger than her husband. Together, they set out for the remote colonial town of Blitar, in the heart of the coffee-growing district.

Like all colonial societies, the Dutch East Indies was characterized by deep divisions between the upper layer of affluent white settlers and the poor indigenous population who worked under them. Herman and Anna Fokker decided to move back to Holland in June 1894 to give their children a proper education. However, the environment of the remote and isolated coffee plantation, with only his sister Catharina Johanna (Toos, born in 1889) and some local Indonesian children as playmates, would leave a deep mark on Tony (or Ton, as he was known to family members).

From very early on, Tony Fokker developed an extraordinarily strong attachment to his mother. She was, and always remained, the cornerstone of his life, a role nobody could take from her. His father, Herman Fokker, was respected but distant, best approached through the intermediary of Tony's mother. Outside the nucleus of his family, Tony Fokker recognized few equals. Although he was probably unconscious of it, the mentality of the colonial environment of his early childhood never left him. The planter's son knew how to use his social and financial status: people could be commandeered, or bought, or, if necessary, both. Fittingly, the first illustrations in the Dutch version of his autobiography, *Flying Dutchman*, depicted his parents and the plantation house in Blitar, the latter a watercolor painting by his mother.

On August 1, 1894, with a month to settle in before the start of the new school year, the family arrived in the stately Dutch town of Haarlem, some fifteen miles west of Amsterdam, a favored retreat for returning colonials. They settled in a roomy, semidetached villa at the Kleine Houtweg, overlooking a park.[3] Endowed with a considerable family fortune and amassed earnings from twenty years as a coffee planter, Herman Fokker retired at the age of forty-three to spend the rest of his years among the circle of ex-colonials.[4]

If education had been a motive for the Fokkers to move back to Holland, young Tony did not live up to expectations. He hated school, every minute of it. His interests lay elsewhere. Having trouble adapting to life within Haarlem's city streets and canals, he would retreat to his childhood sanctuary: the front half of the attic on the third floor of the house where he spent much of his time. There, and all by himself, he would, quite literally, create his own world from scratch. Tony Fokker was an inventive child, with a special gift for making things with his hands. The floor of the attic came to be covered with toy railroad tracks, constructed from strips of metal, on which he ran spring-operated (and later also electric) train sets that he had made himself. As he got older, his self-invented world grew, incorporating wing-warping controls for make-believe airplanes, a little laboratory with Bunsen burners, and miniature steam engines operating on gas he illicitly tapped from a gas pipe leading to the neighboring house. At the age of twelve he was ready for his first real life-size contraption: a small canoe that he managed to build in the attic. His father, who encouraged his handiwork from a distance, was sufficiently impressed to have the boat sent downtown to be painted and readied for Haarlem's main watercourse, the Spaarne River. Boating would remain Fokker's favorite pastime: the first and last constructions of this airplane-builder-to-be were boats.

At school Tony was a bad pupil, despising of his teachers and their efforts "to force a square peg into a round hole."[5] "Classes were boring; teachers stupid; I seemed almost heroically dumb."[6] Always at the bottom of his class, he needed seven years to complete the six-year curriculum of primary school. In high school from September 1903, results were equally bad. He repeatedly skipped classes with the usual variety of excuses, including punctured bicycle tires. In his old school's special sixty-year jubilee newspaper, he later described how he would wait until the caretaker appeared in his office with the list of absentees, then present himself, panting, and beg for his name to be taken from the list, promising to go to class immediately. His name taken off, he would quietly slip away and spend his hours boating on the river Spaarne.[7] Consequently, he had to repeat his second and fourth year, despite elaborate schemes for cheating that he would brag about later in his autobiography.[8] Physical education was the only subject he was really good at. He failed algebra, chemistry, history, and languages, and his command of trigonometry, mechanics, and physics was never more than passable, though he managed to hold his own reasonably well in drawing, draftsmanship, and bookkeeping.[9] At the end of 1908, after a particularly unsatisfactory quarter in his second year in fourth grade (his next to last year), Tony Fokker, now eighteen, dropped out of school.

In keeping with the privileged social class he was born into, Fokker spent the next year pursuing an adolescent dream of quick industrial wealth. During his final months in high school, he had met and befriended a fellow would-be dropout, Frits Cremer. Cremer had an even more afflu- ent background than Fokker's: he was the son of Jacob T. Cremer, former colonial secretary and president of the board of the NHM and of the Deli Maatschappij (an influential Dutch colonial tobacco-trading conglomer- ate dominated by the Cremer family and their close relatives).[10] Like Fokker, Frits Cremer was an obstinate boy with a history of bad behav- ior at school.[11] Together, the two friends set out to develop a puncture- proof tire for automobiles. In those days automobile tires were still made entirely out of rubber, which meant they were prone to spring a leak from sharp objects on the poorly surfaced roads. Fokker and Cremer were de- termined to combat this nuisance and put the rubber tire industry out of business in the process. Instead of the regular rubber tire, they developed a system in which the metal wheels of the car were enclosed by a flat steel belt, fastened to the wheel itself by a array of springs and tension mem- bers. The ideas were Fokker's; Cremer simply lent a hand, moral support, and money.[12] Having been presented with Jacob Cremer's Peugeot car to play with, it took them a whole year and several thousands of their fa- thers' guilders (the Dutch currency unit) to develop a reasonably well- performing mechanism, only to discover that the coveted patent for their invention had already been claimed in France for a similar device.[13]

Fate did not look too kindly upon the young inventor in those days. Having dropped out of school, both Tony and Frits ran the risk of being drafted into the army. Together they cooked up a scheme to secure a med- ical exemption on the basis of having bad feet. In Tony's case this was to no avail. While he was conscripted to fulfill his military service in March 1910, Frits Cremer danced at the Penang Golf Club, one of the stops in a year-long "voyage of discovery" around the world with his brother Marnix and a distant cousin.[14]

If school had been agony for Fokker, military duty was worse. From day one, all his efforts concentrated on getting out. His first attempt at being medically discharged failed. The army's medical examiner hardly took a look at his complaint: legs of uneven length. Tony was sent to the fortress of Naarden, twenty-five miles southeast of Amsterdam, where he was to report with a number of fresh fellow conscripts to No. 9 Com- pany of the 2nd Regiment of the Fortress Artillery. Yet, even on the train to Naarden, Fokker was already scheming for privileges, treating the staff sergeant to expensive cigars to extract information about a possible second medical examination later that week. Although army chow proved passable, the barracks were a far cry from the accustomed environment of an upper-class child. Everything he touched seemed greasy and dirty. His hay sack felt as hard as a brick, and he described the town itself as "a hole to hang oneself."[15]

Fokker later claimed that his continuing pretense of having flat feet accelerated his discharge from the army, but in the end it was more a stroke of good fortune than a result of his abilities to keep up the appearance of a physical handicap. After a few weeks of deliberately falling out of step with the other recruits, he struck and badly sprained and bruised his ankle against the curb of the street when he jumped off a streetcar after visiting his sister Toos in Amsterdam one evening. With this injury he managed to be committed to the hospital for further examination. But the army doctors were in no hurry and did little more than apply wet bandages, put him on a light diet, and order him to stay in bed. The cat-and-mouse game lasted for more than two weeks. Fokker wrote whining letters home to his mother, who answered his cries of anguish by sending him food parcels, mainly dark brown sugar, powdered chocolate, cookies, and tins with sausages he could share with his fellow-sufferers.[16] She sent money, too, and Tony was quick to circulate it, promising the hospital orderly twenty-five guilders—about two week's wages—if he helped him obtain a medical discharge. Fokker had no qualms whatsoever about this bribery, stressing to his mother, "He is First Orderly here, the assistant of the doctor in all operations, and makes good money, so one cannot send him off with a mere handout. Anyway, I will now get out more cheaply than I would in Naarden itself. Patience is the password now."[17]

Indeed, Fokker's careful financial massaging resulted in his accelerated discharge from the army on medical grounds. It was the first time that money helped him escape duty, a strategy that would remain with him. With a dream of pursuing a career in aviation, he returned home triumphantly in the car he had received as a present from his parents on his twentieth birthday.

## BUILDING A DREAM

In 1909–10 aviation was still in its infancy, even though important strides had been taken in the six years that followed the Wright brothers' first successful flights in December 1903. For one thing, the center of heavier-than-air flying activity had shifted from the desolate beach of Kitty Hawk, North Carolina, across the Atlantic to France, Europe's uncontested center of aviation. By the end of the century's first decade, French aeronautical technology, represented by such early aircraft construction firms as Farman, Voisin, Antoinette, Breguet, and Blériot, and French aviators like Henri Farman, Gabriel and Charles Voisin, Louis Paulhan, Jean Conneau, Jules Védrines, and Louis Blériot were competing for prominence with the Wrights. Their achievements, and those of other European aviators, captured the attention of the press and the public's imagination. Fittingly, a Frenchman, Louis Blériot, was the first to conquer the most important of Europe's dividing seaways (which would later attract the highest density of international flights), the English Channel, on July 25, 1909. That summer an aviation craze took hold of Europe. A month after Blériot's remarkable feat, France's champagne capital, Reims, became the center of international aviation for the Grande Semaine d'Aviation de Champagne (the Great Aviation Week of the Champagne district), held from August 22 to 29. A similar international event, the Primero Circuito Aéreo (First Aerial Circuit), was organized in the northern Italian city of Brescia (September 5–13). Such air meets drew participants and spectators from across half of Europe and were fashionable events for high society to be seen at. The Reims week attracted over half a million paying visitors. To entice the competing aviators to explore the very limits of their skills and machines, the organizers offered lavish prizes for various achievements. The unique aerial spectacles had a tremendously powerful effect on those witnessing them from the grandstands, enchanting even such intellectual skeptics as writer Franz Kafka, who started taking classes in mechanical technology upon his return from Brescia to his native Prague.[18] The famed Italian poet Gabrielle d'Annunzio went up with American aviator Glenn Curtis for his first flight in an airplane, which he later envisioned as a biotechnical life form of its own.[19] From then on, France maintained a leading position in aviation, underscored in the annual Paris aeronautical exhibitions, the first of which was held in the Grand Palais from September 25 to October 17, 1909.

In Germany, by contrast, aviation had progressed little, eclipsed by the nation's fascination with the airship.[20] When, on September 26, 1909, Berlin's Johannisthal airfield opened officially with a spectacular flying competition, the only serious competitors were foreigners. The event's most celebrated participant, the debonair "French Englishman," Hubert

Latham (his family was from Le Havre, but the Lathams maintained British passports), got a police fine of 150 marks for making an unapproved and therefore disconcerting "overland" flight from the Tempelhof parade ground in Berlin to Johannisthal.[21] At the Internationale Luftfahrtausstellung (International Aeronautical Exhibition) of October 1909, in Frankfurt, the only German constructor bold enough to present himself, even though the performance of his aircraft could not match that of the French competition, was August Euler. At the time, Euler, who had made his fortune manufacturing bicycles and automobiles, had been attracted to aircraft construction for less than a year. On February 1, 1910, he would receive German pilot license number 1.[22] Of Germany's few other fledgling aircraft constructors, only those with substantial financial backing, like the German Wright Company, sponsored by the Allgemeine Elektrizitäts Gesellschaft (AEG: General Electrical Company); and, to a lesser extent, Rumpler, Aviatik, and Albatros, all in Johannisthal, seemed to have realistic prospects.

Aviation embodied what one historian aptly summarized as "élan, esprit, audace."[23] It combined technology, athleticism, and the thrill of imminent danger and prided itself on unprecedented popularity. Its flying heroes enjoyed an almost godlike status and were treated like royalty. To become an aviator was every boy's dream—Tony Fokker's too, even though he had never seen a real airplane. At twenty, he was living with his parents. He had no diploma, no job, and no clear-cut idea of what to do with his life, only a romantic notion of learning to fly. Herman Fokker would hear none of it, and many a painful hour passed between father and son discussing Tony's future. Finally, Herman Fokker decided to do what his father had done in his own youth: send his son to a school in Germany, hoping that the change of climate would be beneficial and make him more acutely aware of his responsibilities, for Herman Fokker "would not tolerate a loafer around the house."[24] It was decided that Tony should attend the Technikum (Technical School) in Bingen, a small town on the Rhine, some twenty miles west of Mainz. The Technikum advertised itself as a cross between an engineering school and an automobile driving school. It seemed an ideal place to send Tony, especially because its curriculum stressed a practical approach to engineering. Besides, a friend of the family happened to live in Bingen, so Tony would not be completely alone there.

With Thomas (Tom) Reinhold, a boyhood friend who had time on his hands, Tony left on a river steamer for Germany in the last week of July

1910 to see what the school was like and find lodging in Bingen. He felt dejected: the thought of leaving home frightened him to death. In his autobiography, he frankly admitted having never been so discouraged in his life.[25]

Arriving in Bingen, Fokker found the school was little more than a scam to extract substantial sums of money training students for a driver's license. One short tour of the premises convinced him this was a dead end: "The whole information prospectus is humbug, and to go there is throwing money away. . . . All one is taught there is how to drive, and what needs to be known to obtain the official driving license."[26]

That very afternoon, Tony Fokker took the most important decision of his life by committing himself to a singular goal. Despite anxiety about the future and his father's likely negative reaction, he decided to fling caution to the winds and devote all his efforts to achieving the coveted status of aviator. He did so at just the right time: "Had I entered aviation a year or two earlier, I might have broken my neck merely experimenting. Had I waited two years, the technical knowledge then required would have been beyond me."[27]

Fokker and Reinhold boarded a train to Zalbach, near Mainz, and the Erste Deutsche Automobil-Fachschule (First German Automotive Trade School), which was in the process of opening an aviation department pretentiously called the Fachschule für Flugwesen (Trade School for Aeronautics). In 1910, courses in aviation technology were still a novelty, and Tony Fokker was immediately enthused. Describing this institution with the highest praise, Tony implored his father to send a letter to the Bingen Technikum, informing the school of a change in plans, and allow him to register at the Fachschule instead.[28]

Glad to see that his son was at least starting to pursue options of his own, Herman Fokker welcomed the move. Arriving at the school early, Tony discovered that Bruno Büchner, the Bavarian flying instructor and race-car driver, and seven students of aeronautical technology were planning to start the year by visiting the international aeronautical exhibition in Paris. Tony immediately sent his father a letter asking permission to join the group, but to his frustration, the ever-skeptical Herman Fokker would hear none of it.[29]

The aeronautical curriculum of the Fachschule turned out to be even more hands-on than Tony could have hoped for. The school did not own an airplane, but aeronautics students immediately began to build one: a biplane after the standard French Farman design, with a single engine powering a pusher propeller. Around 1910, French constructors were on the cutting edge of aviation technology. To build the airplane, the school had erected a shed at a the small military parade ground at Dotzheim, near Wiesbaden. There, it soon became apparent that Büchner, a self-professed (but unlicensed) pilot, knew little more than his pupils about the field of aeronautics.

In those early years of flight, it took weeks, rather than months, to proceed from construction drawings to an actual flying prototype. When the class had its airplane ready for trials, it quickly became evident when taxiing the aircraft across the airfield that the only aero engine the school had was not powerful enough for takeoff. In other words, the airplane was too heavy, and construction on a lightweight variant began immediately. One of Fokker's fellow students provided a more powerful engine, because the aeronautical section of the Fachschule was running low on money. The practical experience gained from the previous trial proved valuable. Combining a more fragile but lighter construction with a tractor propeller, bigger wings, and a more powerful engine, the second aircraft actually took to the air. Büchner, the pilot, not daring to attempt the dangerous maneuver of turning the airplane in the air, was unable to avoid crashing into a ditch at the edge of the field. The airplane was destroyed, but Büchner, lucky to survive the accident, walked away to pursue a career in aviation, independent of the Fachschule. The aeronautical program did not long outlive its two costly mishaps. Blaming the school for what had happened, Fokker withdrew as a student and demanded to have his money refunded.[30]

By that time the class already had a third airplane under construction, incorporating some of Fokker's ideas about aeronautical design that had originated in his parents' attic. With the performance of the first two machines vividly demonstrating the shortcomings of their designs, Fokker convinced his father to put up 1,500 marks to enable him to buy "his" half-finished monoplane from the Fachschule at the end of October 1910. He wanted to have the machine completed in the workshop of the German airplane constructor Jacob Goedecker at the nearby Gonsenheim flying field of Nieder-Walluf, a mere five miles from Dotzheim.

As it turned out, Fokker's practical design ideas blended well with aerodynamic and construction principles employed by Goedecker, who had established his Flugmaschinen-Werke (Aircraft Works) a year before. Goedecker, a twenty-eight-year-old mechanical engineer from an affluent family background, had first encountered aviation when studying at the

University of Aachen under professor Hugo Junkers (whom we will mention further on in the book). At Gonsenheim, Goedecker built aircraft based on designs originally drawn for the air-minded Austrian textile entrepreneur Igo Etrich in 1909. Etrich's eye-catching Taube (Dove), named after its dovelike wing shape, an airplane with extremely stable flying characteristics, was quickly conquering the German aeronautical field. Initially Etrich had applied for a patent for his Taube and sold exclusive license-production rights to the Johannisthal constructor Edmund Rumpler. The German Patent Office, however, ruled that the Taube design expanded on structural principles already published in 1897 and refused to issue the patent, opening up the possibility of copying and improving the Taube. Goedecker was one of a number of small manufacturers building on that option.[31]

Moving to Goedecker's workshop was Tony Fokker's first step on the road to becoming an aircraft constructor in his own right. But his more immediate goal was finding money to buy an aero engine powerful enough for his machine. For this Fokker managed to enlist a man of independent means, the retired *Oberleutnant* of the German army, Franz von Daum. Von Daum, in his early fifties, had been one of the students in the Dotzheim program and was eager to learn to fly. Putting up the money to buy an engine for Fokker's aircraft seemed a shortcut to achieving this goal.

Late in October, the Fokker–Von Daum monoplane was completed. Wary of the limited size of the flying field, Fokker and Von Daum transported it to the spacious airship grounds of Baden-Baden.[32] It was a fitting place for aeronautical experiments, not only because of abundant space, but also because everything to do with airships, then generally considered the ultimate symbol of German technological prowess and international ambitions, commanded attention.[33] Flying from one of the Zeppelin stations, Fokker and Von Daum were more likely to attract public notice and thereby obtain funds for further experiments. Not until mid-December was everything more or less working properly and the airplane ready for flight testing. The crucial problem was how to control the machine while it was still on the ground, because the original prototype, like most early airplanes, did not have a steering mechanism other than twisting the wingtips in flight. This maneuver, known as wing-warping, had been developed by the Wright brothers in the early years of the century.

Shortly before Christmas of 1910, Fokker, treating the machine as if it were entirely his own and keeping Von Daum away from the controls, managed to fly a few short hops of a hundred yards or so. He was elated: his airplane could actually fly! In three short months, he had gone from being a complete outsider to achieving the much-coveted status of aviator—and with a machine of his own design, no less. However, this incredibly rapid success came at a price. Sweaty work on the airplane, muscle-straining efforts to start the engine, and exposure to icy December winds on the ground and at speeds of up to 40 mph in the air caused Fokker to contract pneumonia. Running a high temperature, he went home to recuperate over Christmas.

That left Franz von Daum, who had actually put up most of the money, in charge. Understandably, after Fokker's departure, he took the controls himself and crashed the airplane into the only tree on the whole field. Notified by telegram, a furious Fokker returned to Baden-Baden, only to discover that the machine was beyond repair.

An improved version of the airplane was built in Goedecker's workshop at Gonsenheim in a mere six weeks from completion of the drawings—a record, wrote Fokker in a letter home.[34] Fokker moved to Nieder-Walluf and lived in a small cottage near the airfield with another young Dutchman eager to be taught to fly, Bernard de Waal. Improved through experience and Goedecker's know-how, the new aircraft incorporated better lateral and vertical controls and a strengthened undercarriage. Goedecker also manufactured the airscrew. Early in May 1911, the new monoplane was ready for testing, and on May 5, Fokker dared to perform the dreaded maneuver of turning in flight, circling the airfield three times.[35] Eleven days later, he felt confident enough to face the examiners of the Fédération Aéronautique Internationale and pass his license examination by flying two figure-eights in his own airplane. On June 7, 1911, he was issued German license number 88.[36] With it, Fokker went to work for Goedecker as a flight instructor in Goedecker's newly opened flying school, and as a demonstration pilot. Only three days later, Fokker was confident enough of his own abilities that he agreed to be Goedecker's demonstration pilot at the Deutscher Zuverlässigkeitsflug am Oberrhein (German Upper Rhine Reliability Flight). And within a few months, he flew one of Goedecker's aircraft in the autumn maneuvers of the army in the Mainz region, logging a flight of thirty miles in a single day. As the airplane's constructor, Goedecker was laureled for the achievement.[37]

After Fokker earned his flying license, his partner, Franz von Daum, demanded access to the airplane as well to train for his pilot's license. How-

ever, early in June, after a few initial hops, Von Daum crashed again while
landing. Lucky to survive the accident, but unscratched, he renounced flying and accepted Fokker's offer to buy him out for 1,200 marks. Again
the money came from Herman Fokker, who would, according to Tony's
autobiography, invest a grand total of about 183,000 marks to establish
his son's flying career.[38] More than anything, it was the largesse of
Fokker's family that made his career in aviation possible.

That summer Tom Reinhold and Frits Cremer, the latter back from his
round-the-world trip and keen to be taught to fly on their friend's machine, joined Fokker at Goedecker's workshop in Gonsenheim. In August, Fokker's new, and third, monoplane was ready, powered by the only
engine Fokker owned. It was completed in time to transport it to Holland
by river barge toward the end of the month. Some months previous
Fokker had agreed to his father's express request to give a flying demonstration in Haarlem with his new aircraft, now more or less officially
called *Spider* after the myriad piano wires supporting the airframe. It being the custom to celebrate the queen's birthday, August 31, with various
festivities in every Dutch town, the Haarlem organizing committee, of
which Herman Fokker was a member, had invited Tony to come as the
year's special attraction. The committee had arranged to transport his airplane and prepare a rudimentary flying field outside town. Immensely
proud to be a "hometown boy made good," Tony immediately accepted.
Although the small pasture provided by the committee was responsible
for some hair-raising moments at takeoff and landing, the high-school
dropout was enormously gratified to be honored as one of Haarlem's outstanding sons. "This reception I will always remember as the biggest satisfaction I ever got out of aviation."[39] Cashing in on the moment, Tony
obtained promises from his father for further financial support to a maximum of 50,000 guilders.[40]

Overshadowed by Goedecker, Tony's own business at Gonsenheim
remained stationary. Between August and December, he was forced to
withdraw 11,000 marks from the credit his father had provided.[41] Late in
November, he decided to move to Johannisthal, the center of aeronautical activity in Germany, southeast of Berlin. With his friend Frits Cremer
and Bernard de Waal as flight instructors, Fokker set up his own flying
school there and hired a professional mechanic, August Nischwitz, who
had been in Goedecker's employ.

Fokker's move had been in the making from the start of his association
with Goedecker. Building two types of airplanes in one workshop had

proved difficult because Fokker felt that Goedecker was giving his own designs preferential treatment on the work floor. Another factor was the gravitational pull of Berlin. Although he and Goedecker got along well, Fokker gradually became convinced that the Berlin entourage offered better business prospects.

In Johannisthal, firms like Rumpler, Albatros, and the German subsidiary of Wright enjoyed the growing patronage of the army, the result of recommendations for the training of military pilots from the chief of the General Staff, Gen. Helmuth von Moltke, in February 1910. It has to be remembered, however, that Germany was not at the forefront of aeronautical development in 1910–11. It was strangely contradictory that the rapid development of Germany in the political, economic, financial, and general technological spheres after 1871, and the buildup of the German navy, was not mirrored by a clear policy of stimulating the modernization of the army until a few years before 1914.[42] Lacking sizable army support, the Germans were late starters in the field of powered flight.

Some of the rapidity with which Fokker came to the fore of the aeronautical scene in Germany in the months following his successful flights at Gonsenheim went back to the slowness of aeronautical development there compared to that in France. Coordinated efforts only materialized in 1912 as a result of the Nationale Flug Spende (NFS: National Air Subscription), when Prince Heinrich of Prussia, the kaiser's brother, who had just learned to fly at Euler's flying school, collected a grand total of over seven million marks in public support for a large pilot training program. Because aircraft construction and flying schools were commonly combined in a single company, the German aviation industry benefited from the generous NFS payments for each newly licensed pilot. But until the impetus of the NFS in aeronautical development manifested in a rising demand for new aircraft, German aviation developed slowly, enabling newcomers like Fokker to catch up quickly.

In Johannisthal, Fokker hoped to escape his marginalization in aeronautical developments. He also aspired to catch the attention of the German army, a growing source of funds, and jump on the bandwagon of pilot training, the income from which he needed desperately. Planning for the move, which meant severing some of the links with Goedecker, took until early December 1911. Goedecker continued building airframes to Fokker's design in 1912 and 1913—twenty-five in all—to undergo final assembly in Fokker's new workshop at Johannisthal. Again, things did not proceed as Fokker hoped. Although Helmuth von Moltke proposed a rapid expansion of airplane production for army use in the wake of the Franco-German Morocco crisis of 1911, the War Ministry was not forthcoming with the necessary funding to support a larger aviation industry.[43]

Despite some improvements in financial backing through a partnership with Berlin aviation enthusiast Hans Haller, 1912 was an extremely tough business year for Fokker. In *Flying Dutchman,* he noted having to ask his father repeatedly for another 4,000 or 5,000 marks to keep his venture afloat.[44] Other than nebulous future prospects and high enthusiasm, Fokker had nothing to show in business terms. Consequently, in his cumbersome prose style, he was constantly searching for the right nuance of flattery that would release the next parental loan.

> Everything that I have achieved, and not only that, but that which has brought me the knowledge and basis to stand on my own feet at all times, be there, however, no high-school diploma, and has made for myself a reputation that is worth more to me than a diploma, all of that is thanks to the fact that you have given me the confidence of a first, and thereafter second loan, for which I am deeply indebted to you. . . . Believe me, I have had difficult times, and still have them, but hope to be able to overcome these time and time again. I struggle with the means that I have, and hope you will replenish these means.[45]

## BECOMING A BUSINESS

Tony Fokker officially entered his venture in the Charlottenburg business register under the name of Fokker Aviatik on February 22, 1912, with a working capital of 20,000 marks, but the company operated as Fokker Aeroplanbau in Shed Number Six in Johannisthal.[46] Orders were extremely slow coming in, and the flying school attracted few new pupils because Fokker had yet to carve a name for himself amid the established competition of AEG, Rumpler, Albatros, and Luft Verkehrs Gesellschaft (LVG: Air Transport Company). Thus, Fokker concentrated on giving flying demonstrations and improving the Spider, fitting a fuselage nacelle, replacing the original bracing wires with steel cables, and installing a more powerful engine. Such changes brought Fokker's feeble Spider in line with the more robust-looking Taube derivatives that Rumpler and Albatros were building across the airfield. Fokker prepared one of his airplanes to be easily transported behind an automobile, anticipating that he would be asked to demonstrate the Spider before the army at the Döberitz airfield in July. Yet, four months after the founding of Fokker Aviatik, business still remained at ground level.

With time, however, Fokker's exceptional piloting ability, evident in numerous demonstrations he gave at Johannisthal, attracted high-level attention and more pupils. In one demonstration, he managed to pack three passengers into his small airplane, impressing the Bavarian prince-regent, who was a spectator. Ten days later, on May 24, 1912, flying with a passenger in bumpy weather, the near-inevitable happened: like so many of the early aviators, Fokker fell from the sky when one of the wing bracing wires snapped, causing the wing to collapse when he attempted to make an emergency landing. Miraculously, he survived with no more than a few broken ribs, though he slipped into unconsciousness for a few hours; his passenger, an army lieutenant, died.[47] Obviously, the airplane's structure needed strengthening. At Haller's suggestion, Fokker met with Reinhold Platz, a twenty-six-year-old mechanic (born January 16, 1886) and an expert in oxy-acetylene welding. A relatively new process in those days, oxy-acetylene welding was just beginning to find its way into the aeronautical world after being introduced in Germany by the Cologne-based constructor Arthur Delfosse in 1911. Fokker believed that, if performed properly, welding could be useful in airplane construction because it made clean, lightweight metal joints and offered great inherent strength without the need for periodic checking. Despite the sorry state of the company's finances, Platz joined Fokker toward the end of May 1912. At that time the company employed some twenty people and owned four airplanes.[48]

Figuring that his chances of obtaining a German army contract had been seriously reduced by the fatality suffered in the crash, and with blunt threats from his father that unless he received some orders quickly, he would stop financing the bankruptcy-bound endeavor "even if you stand on your head,"[49] Fokker was forced to search far and wide for orders. Early in 1912 his attention had been drawn to Russia, when two Russians, an engineer called Grünberg, and a noblewoman, Baroness Leitner, had engaged him for flying lessons. From Grünberg, Fokker gathered that the Russian army was organizing an international flying competition, which would be followed by an order for six machines from the winning competitor. Contestants were to receive 2,000 marks to compensate them for the cost of transporting their machines and equipment to Saint Petersburg. Among those in the competition were the Russian constructor Igor Sikorsky and a fellow daredevil pilot from Johannisthal, the Russian aviator Wsewolod Abramowitch, representing the German Wright Company.

With yet another Dutch arrival at his Johannisthal operation, Jan Hilgers (the first Dutchman to fly in Holland, on July 29, 1910), Fokker set off for Saint Petersburg in August 1912. The trip turned out to be an expensive lesson in the art of doing business. Despite the qualities of his aircraft, Fokker got nowhere. He knew how to oil the working parts of his machines, but he still had to learn how to oil the hinges of the military-procurement bureaucracy. And though Fokker later made a case against the omnipresent bribery he encountered on his Russian trip, his dismay was, in retrospect, probably more inspired by frustration over the fact that he lacked the means of inducement than by the practice itself.[50] Although the Russian excursion caused him to suffer yet another acute financial crisis, it ended on a bright note. In Saint Petersburg, Tony Fokker was introduced to the nineteen-year-old Russian aviator, Ljuba Galantschikova, who spoke fluent German and French. Fokker later said, with some degree of truthfulness, that in those days, "Aviation was the whole of my life. To interest me a woman would have had to be a flyer herself. I was working day and night and had no time for play."[51] Tony Fokker fell head over heels in love with the vivacious and attractive Galantschikova, imploring her to come to Berlin with him so that he might "teach her how to really steer my plane."[52] Ultimately, however, she was not as interested in the constructor as in his airplanes. Fokker later confessed in the Dutch (but not the American) version of *Flying Dutchman* that aviation made up all the conversation he was capable of, despite his being in love.[53] She did come to Johannisthal with him, though, and bought his high-performance (military) Spider, in which she established a world altitude record for women that fall (2,200 m, or about 7,200 ft.). In the following months, Fokker hired her as a demonstration pilot to cash in on the extra attention his aircraft received from being flown by a woman. He liked to parade her as his "Russian girlfriend from St. Petersburg," which she allowed him to do, though she soon lost interest in the one-track-minded Dutchman with his high-pitched voice. With surprising frankness, Fokker revealed to his Dutch and German readers: "I was angry, I was jealous, and I was too proud to show my feelings. If a woman even looked at another man I did anything rather than admit she still interested me. Even though I would feel absolutely miserable, I would not show it to her."[54]

Instead, Ljuba had a brief but passionate fling with another Johannisthal aviator, Gustav Adolf Michaelis, whom Fokker described as a handsome bon vivant with a 100 hp car and a man who knew how to deal with women. Michaelis crashed on May 27, 1913, and died of his wounds five days later. The loss did not leave Galantschikova heartbroken for long. She found comfort in the arms of the French aviator Léon Letord, and they flew off to Paris (nonstop) on July 27 in Letord's sleek Morane-Saulnier, leaving Tony utterly depressed and in a self-destructive frame of mind.[55] It took Fokker six weeks to drown his sorrow by plunging into the notorious nightlife of Berlin.

> For the first time in my life I learned something about women. I visited the nightclubs, and met a kind of women there that I did not know existed. I found them enchanting, merry, and ravishing. Yet, I found it impossible to treat them in the same way other aviators did, so uncaring, and with disregard for their own true feelings. I liked hearing them speak about their life, and studied their character. More than once I have tried to help a girl whom I admired find different, more worthy employment.[56]

Tony Fokker was enough of an entrepreneur to file for patents on his automatically stable Spider and various other aeronautical inventions, but he was still scraping the bottom of the barrel financially. By mid-September 1912, largely because of his Russian expedition, he had spent the full 50,000 guilders his father had set aside to invest in his endeavors. Despite the stimulus of the NFS, orders were still not forthcoming. Hard-pressed for cash, Fokker thought his lucky day had arrived when he was approached by a group of corporate bankers with ambitious proposals to invest in his fledgling business. He was easily convinced that their promises of hundreds of thousands of marks were solid and signed a contract relinquishing the aeronautical patents he had taken out since 1911 and the control over his venture. In return, the bankers promised to make 20,000 marks available as short-term credit. But there was a catch: Fokker could have the money if and when he secured a contract for the delivery of aircraft to the German army. When that contract did not materialize, the financiers backed out, leaving him virtually stranded.[57] For the first time, his father, approaching the limits of the resources he had put aside for his son's future, refused to provide financial relief. Despite Tony's repeated arguments that his father stood to lose all his investments if he pulled the plug now, Herman Fokker announced he had come to the end of the road as far as money was concerned:

If you understood anything about business you would know that all businesses can go wrong. If I would now be so stupid to keep furnishing you until I would have invested, for example, half or three-quarters of my estate in your venture, and your business went belly-up, then I would not only be forced to go hungry or find myself a job in my old age, but I also have a wife and daughter who would be dragged down in the demise. I would act like a criminal if I were to risk their future for your pleasure. . . . Perhaps I would have risked <u>more</u> than 50,000 guilders if I would have received proper monthly accounts of the capital used, which it was your <u>duty</u> to provide, but now I have been forced to set a maximum. . . . Now I have received notice from the bank that you have used up the <u>maximum,</u> from which obviously follows . . . that the bank will not provide any more money. And neither will I, of course. This is what I meant when I stated before that you never seem to understand what the word maximum means.[58]

Given his lack of business expertise, Tony Fokker had few options. At this point the advantages of his family background once more saved his career. One of his father's older brothers, uncle Eduard Fokker, who had been receptive to the winged dreams of his young nephew, agreed to provide short-term credit of 20,000 marks—the same amount Fokker had counted on receiving from the German bankers. But for long-term financing, Tony continued to make renewed appeals to his father. When Herman Fokker kept giving him the cold shoulder, his requests soon turned into wailing. Appealing to his mother, he sent home a photograph with the caption, "The following image depicts . . . Tony at Johannisthal, skinny because of worries and hunger. Postcard received. Adio [sic]. Regards to Pa and Toos."[59]

In the end Fokker did manage to extract more funding from his father, who must have recognized his son's talents, before his uncle's 20,000 guilders ran out. Nevertheless, Herman Fokker made it clear that the limits of his parental benevolence had truly been reached and that he had invested all the capital he could spare in his son's aeronautical career. In the future Tony would really have to look elsewhere for venture capital. As a result, Tony's life remained a struggle in 1913.

To a considerable degree, his continued financial hardship was the result of a bizarre scheme that his Dutch associate, Hilgers, had proposed and which Tony Fokker agreed to. On December 20, 1912, they signed a contract under which Fokker supplied two Spiders with which Hilgers, who had also been born on Java, was to make a demonstration tour in the Dutch East Indies and then sell the airplanes there. Fokker provided the aircraft free of charge, while Hilgers defrayed the cost of transport and demonstration. The proceeds were to be split fifty-fifty. The venture was a financial disaster for both. Hilgers departed with the aircraft and never returned, setting Fokker back another 40,000 guilders.[60]

Fortunately, the Fokker flying school started to attract customers, both civilian and military. Despite a precarious cash flow, the business expanded slowly throughout 1913, with Frits Cremer and Bernard de Waal taking care of most of the money-making flight instruction. In July 1913 Fokker Aviatik/Fokker Aeroplanbau finally struck gold by landing an army contract for four airplanes that could be rapidly assembled and disassembled for road transport, an idea that Fokker had been working on for over a year. A month later, the army ordered a second batch of six aircraft. The purchase totaled 299,800 marks, an unprecedented sum for Fokker.[61]

Such sales reflected rising army acquisition of aircraft. If the army had bought only 28 airplanes in 1911, its procurement figures expanded fivefold the next year and more than tripled to 461 in 1913.[62] Still, the army preferred to buy from well-established companies, and further contracts failed to materialize for Fokker. Army procurement of monoplanes focused on two basic types of aircraft: Type A comprised slow but stable monoplanes almost exclusively of the Taube configuration, while Type B included more maneuverable biplanes. About equal numbers of each type were produced in 1912. Both relied primarily on one type of power plant: 100–125 hp six-cylinder in-line engines built by Daimler or Mercedes. The army had decided to standardize on these two types for its first-line aircraft in an effort to arrive at a rational, cost-effective army air wing. The Taube aircraft were inherently stable in the air, extremely easy to fly, and therefore thought ideal for officers, who would not unnecessarily risk their lives. That these characteristics made the Taube sluggish on the controls and difficult to turn mattered little. Production was spread out over no fewer than ten manufacturers, including Fokker's direct Johannisthal competitors Rumpler, Albatros, and LVG, and totaled around five hundred aircraft.[63] Though Fokker's Spider-type aircraft were also automatically stable, these "foreign" designs played only a minor role in German planning, and Fokker Aeroplanbau was directed to move away from Johannisthal to make room for the expansion of the more powerful firms.

The idea of moving to the town of Schwerin, in Mecklenburg, some 130 miles north of Berlin, was first put to Fokker by Capt. Franz Geerdtz of the General Staff, whose job it was to reorganize aircraft production to meet the army's needs in the event of war. Possibly to compensate for further marginalization, Fokker was offered very favorable conditions. His move fit well with the general pattern of Germany's accelerated military buildup in 1913.[64] With support from the city of Schwerin, a corporation had purchased fifty-three hectares of land close to Ostorfer Lake, at Görries, a few miles south of Schwerin, for the purpose of constructing an airfield. Fokker was offered the use of hangars and some other buildings for his flying school under a favorable long-term lease with an option to buy later. That autumn, the city of Schwerin erected a 66 by 85 foot (20 by 26 m) wooden workshop for the construction of seaplanes a short distance away, on the shore of Ostorfer Lake, with access to the water via a ramp.

In early June of 1913 the military wing of the Fokker Flying School was the first part of Fokker's operation to move north, taking a handful of airplanes and their pupils with them. Tony Fokker was pleased with the relocation, which he interpreted as a sign of support from the army. It appeared to him that he might finally be heading for a breakthrough in the aviation business. He appointed his chief mechanic, August Nischwitz, as works manager in Schwerin until Fokker also moved his construction activities there on October 1. To make optimum use of the facilities, he needed to expand the scope and scale of his business. Once more he traveled to Holland asking for support. A conference was organized with "the uncles" (as Herman Fokker's five brothers were usually referred to), Jacob Cremer, and a few interested friends of the family. Between them, they promised funding of 300,000 marks if Fokker could land a substantial order from the German military.[65] But further sizable orders from the German army, to whom Fokker kept submitting new prototypes, still did not materialize. Despite the fair number of pupils the Nationale Flug Spende brought to the Fokker Flying School, the future of his outfit as an aircraft construction company looked as bleak as ever at the end of 1913.

To shore up the promised new funding, Tony Fokker badly needed to come up with a superior aircraft. Never conventional in his approaches, he decided to take a big, though covert, step ahead. In late December 1913, Fokker and Haller boarded a train for Paris, where Tony had learned of a cheap, damaged Morane-Saulnier Type H for sale at a bargain price of 500 marks. The Morane-Saulnier, powered by a lightweight Gnôme air-cooled rotary engine (instead of the heavier in-line water-cooled engines German manufacturers used), was known to have flying characteristics superior to those of any of the German-designed aircraft at that time. Fokker bought the airplane, which was quietly transported to Schwerin and put in a shed at Görries to be taken apart and reassembled so Fokker could copy its design characteristics and incorporate some of the latest French design technology in his own aircraft. Such imitation, reeking of industrial espionage, was not uncommon at that time.

Taking the Morane-Saulnier monoplane as the basis for a new model, Tony Fokker and his new chief designer, Martin Kreutzer, built an airplane that resembled the Morane-Saulnier but which was lighter and sturdier because it had a fuselage of welded steel tubes, instead of wood. Completed in April 1914, the Fokker M.5 (M stood for *Military;* the aircraft was later called the M.5L, for *Long span*) was a shoulder-wing single-seater, powered by an 80 hp Gnôme rotary engine, manufactured under license in Germany by the Oberursel engine factory. It gave the aircraft a maximum speed of about 80 mph. Fokker and Kreutzer had chosen to keep the design as simple as possible. The 23.9-foot fuselage was box-shaped, the steel tubing covered by easy-to-repair fabric. Forward of the cockpit, the fuselage was covered with metal and ended in a horseshoe-shaped fairing over the engine, also copied from the Morane. The 31.4-foot wings were made up of a glued-together wooden structure, covered with fabric. Lateral movement was controlled by a system of twelve wing-warping cables operated over a tall pylon protruding upward from the fuselage. On the basis of the M.5 airframe, Fokker subsequently constructed a two-seat trainer and artillery-spotter version, the M.8; and a more maneuverable 28-foot short-span variant, the M.5K.

Fokker took the new monoplane on a tour to demonstrate its aerobatic qualities, performing the same maneuver that had stunned Berlin some months earlier, when the French pilot Adolphe Pégoud thrilled spectators by flying a loop at Johannisthal. The spectacular aerobatics and the handling characteristics of the new airplane caused quite a stir in German military circles. On June 4, 1914, the chief-of-staff, Gen. Erich von Falkenhayn, personally inspected the wonder plane. Tony Fokker was decorated with a laurel wreath for his spectacular feats.[66] In the nick of time, he had pushed himself into the limelight of the German military buildup preceding World War I.

On June 28, 1914, Gavrilo Princip, a radical Serbian nationalist student, assassinated the Austrian crown prince, Franz-Ferdinand, when the prince visited the Bosnian capital, Sarajevo. The assassination ignited a controversy between Austria-Hungary and Serbia that went back to

1878, when Austria successfully invaded and occupied Bosnia and Herze-
govina and replaced the Ottoman Turks there as the occupying power.
The act was viewed with chagrin by Serbian nationalists striving for a nation of their own as the nucleus of a "Greater Serbia" uniting all Slavic
peoples living in the Balkans. It brought them at loggerheads with
the Austrians. In the conflict between Serbia and Austria that evolved, the
Austrians, who enjoyed the comforts of military superiority, formally an-
nexed Bosnia-Herzegovina in 1908. The Serbs then sought to achieve
their objective by supporting a secret society founded in 1911: the Society of Death, also known as the Black Hand, of which Princip was a devoted member.

For Fokker, the rapid stream of events that followed Princip's shots,
bringing to life a system of military alliances between the major European
powers, resulted in increased orders. The army ordered twenty-four of his
M.5 monoplanes, and twenty M.8 two-seaters were virtually ordered off
the drawing board. The navy even went as far as to confiscate three machines and later saw to it that some Fokker monoplanes and biplanes
were diverted to its own Air Service. Fokker also received orders from the
Austro-Hungarian government for military aircraft. Elated that recognition and success had suddenly blown away years of anxiety, Fokker
agreed to every order he could get, even though, like the other German
producers, he was in no way equipped to deal with them.[67]

Additional capital was now immediately necessary. On July 7, 1914,
his long-awaited limited-liability company, now formally named Fokker
Aeroplanbau GmbH, was finally incorporated in Schwerin. Uncle Eduard
Fokker committed himself to a maximum of 60,000 marks, while Uncle
Anthony Sr. and his wife, Aunt Susanna Fokker-der Kinderen, signed up
for 50,000 marks. Frits Cremer's father, Jacob Cremer, pledged 40,000,
while Frits himself invested 35,000. Two friends of the family from The
Hague, Constantijn van Schmid and Everwijn Levert, put up 15,000 and
10,000, respectively, and Tony himself suddenly found himself in the position to participate with up to 90,000 marks, which were to come his
way as payments and advances from the military. This brought the nominal capital up to 300,000 marks. Fokker's Dutch investors paid 25 percent of the 300,000 marks upon signature of the memorandum of association, and Fokker himself brought forward his share in the form of fixed
and portable assets.[68] Tony Fokker, who only a few years previously had
dropped out of high school, had finally and suddenly become big business, and a business, moreover, with government support.

# 2

## THE FORTUNES OF WAR

In the German strategic planning for the eventuality of war, aerial operations took up a position of some importance. Because of the prominence of airships in the fascination with flight that was so characteristic of German officialdom and public alike, it was envisaged that Zeppelin airships, operating at altitudes too great to be reached by ground and air defenses, would be employed for strategic bombing attacks, primarily against Britain.[1] Since 1912 the General Staff's plans had provided for the use of aircraft in battlefield operations. Their primary role would be limited to short-range aerial reconnaissance and communications. As early as 1907, the possible deployment of airplanes for such purposes had been pointed out to German army planners by Hermann von der Lieth Thomsen, a captain of the General Staff, who was to play a key role in the formation of the German Flying Corps. A year later the German military had become interested in acquiring a Wright Flyer for evaluation, but negotiations stalled on the excessive price ($200,000) the Wrights demanded for their machine and the use of their patents. A limited-scale pilot training program started at Johannisthal in 1910, using German-built aircraft, most of them of the Taube design. But not until 1913 were substantial numbers of military planes ordered as a result of the NFS. Orders were placed so that larger and well-funded companies like Albatros, Aviatik,

LVG, Rumpler, AEG, and Gotha would be encouraged to adopt mechanization and series production of slow but stable reconnaissance airplanes, powered by standardized Daimler, Benz, or Argus engines.[2] As a consequence, Tony Fokker had remained on the sidelines of German aeronautical developments, surviving on small orders that showed the army was interested in his designs, even though it would not adopt them for general use. But copying the Morane-Saulnier changed his luck.

As a whole, the German aircraft industry, still in an embryo stage when hostilities broke out in August 1914, had produced only limited numbers of military airplanes. The total German frontline strength amounted to about 250 reconnaissance aircraft, no more than a handful of them Fokkers. Against these, Germany's enemies assembled a total of some 390 aircraft, 240 in the west (about 140 French and 100 British), and 150 Russian aircraft in the east.[3] Because aerial observation of army movements endangered the success of such maneuvers, keeping enemy aircraft from overflying the front line was soon recognized as a military necessity. The use of airplanes brought about a fundamental change in modern warfare: if surprise was to be gained in an attack, armies had to move at night or under the cover of low clouds and bad weather. The First World War thus moved into the air. More important than sheer numbers was the fact that in British, and particularly in French, military circles, the anticipated role of aircraft soon progressed beyond observation. There was a growing recognition in Paris that air superiority was a necessary element in modern warfare.[4] From the beginning of October 1914, the French started arming their airplanes (rather, their aviators) toward the development of a military aviation division as a weapon in its own right. They began using their airplanes to chase German reconnaissance aircraft from the skies over the Western Front. The first recorded aerial victory occurred on October 5, 1914, when a French Voisin pusher-biplane with a machine gun mounted in the front cockpit shot down a German observation plane.[5]

Marred by organizational, logistical, and industrial problems, the Germans were relatively slow in responding to this changed perception of aerial operations. Initially, they focused on procuring enough machines to offset operational losses at the front, estimated to run as high as 40 percent in August 1914, when frontline pilots were noticed hanging around aircraft factories waiting for priority deliveries. At meetings of military, industrial, and political representatives, it was agreed to boost aircraft production to about 200 aircraft a month from mid-September, even though the aero-engine industry could not guarantee more than 170 engines.[6]

When German military advances on the Western Front slowed to a more or less stable trench war after the battle on the Marne and cavalry reconnaissance became impossible, the airplane's role in gathering information on enemy movements became pivotal. The ensuing necessity to expand aircraft production rapidly opened undreamt-of opportunities for Fokker. Though the vast majority of German frontline aircraft were built by the big firms like LVG, Aviatik, and Albatros, Fokker's adapted Morane-design M-series, the only lightweight airplane available in Germany at the time, found its niche. The army was willing to buy, and would even confiscate, every aircraft it could lay its hands on. Fokker's designs, powered by the German version of the French lightweight (air-cooled) Gnôme rotary engine, built under license at the factory of the Oberursel Motoren Werke in the town of that name near Frankfurt, had an additional advantage. A production run of Fokker M-planes would not depend on the availability of Daimler, Benz, and Argus in-line water-cooled engines, which were already in dramatically short supply.

In the heat of the moment, Tony Fokker leaped at this opportunity. Realizing that his breakthrough was at hand, he made delivery offers to the German army and navy, and to a delegation from the Austro-Hungarian air service, the Luftfahrttruppe, in September 1914.[7] His willingness to commit himself went far beyond the capacity of his factory, staffed as it was by no more than fifty-five people in August 1914.[8] To boost production quickly, Fokker faced the choice of taking on more personnel; rationalizing and mechanizing part of his production; or investing in a second production facility. Because there was no way of telling how long the war would last, he chose the first of these possibilities, doubling his work force by the end of the year.[9] After years of scraping the bottom of the barrel in aircraft construction, it was wise to be cautious. In meeting increased demand with a larger work force and postponing investments, Fokker followed the normal practice in the still comparatively fragile German aircraft industry.

Curiously, for a man who had staked his career and destiny on a gamble, his policy in later years was marked by a conservative approach. In Fokker's later postwar factories, when he could easily have made other choices, he continued to meet increased demand with more workers instead of establishing mass production methods and acquiring more fixed assets in the form of modern tooling and machinery. In the early 1930s, this reluctance to invest would have serious repercussions.

Though the growing number of employees at Schwerin did their utmost to increase output, delivering the first of a series of forty M.8 two-seat observation planes to the army in October, no more than twenty-three M.8 aircraft were produced before the end of 1914. Fokker's productivity per worker fell well short of that of the best-run companies like Aviatik and LVG, though it was still substantially better than that of the much larger Rumpler firm, where production continued to be extremely inefficient.[10] Besides the M.8, Fokker also built a small number of single-seat M.5 monoplanes. But while the flying characteristics of the M.5 and M.8 met with enthusiasm from frontline pilots, the Fokkers were plagued by a series of troubles, most of them engine-related. Fokker could do little about this, for he was not at liberty to chose the best available engines for his aircraft. Engine allocation was governed by the Inspektion der Fliegertruppe (known by its acronym, Idflieg), which controlled all landplane and aero-engine production. The early Oberursel engines were notably untrustworthy, as several pilots discovered to their misfortune, and Fokker pilots were advised not to venture across enemy lines. Despite his enthusiasm about the Fokker's flying characteristics, one pilot noted as late as August 1915: "I will not fly this Fokker fighter. I find there have been too many accidents with it lately."[11] A special investigative commission was sent to the front by Idflieg to determine the cause of the high accident rate. Sorting out what caused frequent overheating of the engine cylinders took Idflieg and the Oberursel engineering staff a long time. Eventually, the problem was traced to the buildup of castor oil residue in the cylinders. The rotary engines had to be cleaned after every flight—something the Germans did not do early on. Careful servicing solved the problem.

Army orders for Fokker M-planes finally put Fokker Aeroplanbau on a more secure financial footing. While most of the German aircraft industry complained bitterly about low aircraft prices and the upwardly spiraling costs of raw materials and of organizing mass production, Fokker rejoiced in the increased cash flow. He offered to buy out his father, uncles, and the other befriended Dutch shareholders while the going was good. The German military authorities welcomed this move because it did away with the non-German capital tied up in Fokker's enterprise. A financial arrangement was negotiated with Bankhaus Strauss & Co. in Karlsruhe. The buyout offer, rather unexpected after years of nothing but

losses in a venture that appeared to be stalling, was gladly accepted in
Holland, and Fokker was even able to repay his long-time Berlin fi-
nancier, Hans Haller. In *Flying Dutchman,* Fokker would proudly remark
that he completely owned his business from then on.[12] As a result of this
financial restructuring, Fokker Aeroplanbau GmbH was dissolved on
January 13, 1915, and a new firm was founded in Schwerin, Fokker
Flugzeugwerke GmbH, with Tony Fokker as its sole proprietor.[13]

With increased orders for the Fokker M-planes and growing recognition of their flying characteristics by frontline pilots (Oswald Boelcke,
who later became an ace, wrote to his parents that his Fokker was his
"best Christmas present," in which he took "a childish pleasure"),[14] the
German authorities became concerned about Fokker's status as a foreigner. They "advised" him that it would be in his best interest as an aircraft producer to adopt the German nationality. Putting his mouth where
the money was, Fokker readily filed his application for German citizenship on December 22, 1914. It was accompanied by a letter imploring the
military authorities to process his request as speedily as possible so he
would not lose any time over the matter after his return from a short visit
to Holland over Christmas to settle the financial arrangements with his
family.[15] The request was granted, and his new nationality and passport
came in the form of a birthday present on April 6, 1915.[16]

In the months after his return from the Dutch trip, Fokker and his staff
set about experimenting with various new designs for improved M-type
aircraft. Fokker approached design and construction with what was primarily an empirical method. Of course, design calculations were made,
but as a general rule computations followed practical experiments, not
the other way around. Aircraft were not conceived primarily as a result
of design and arithmetic on the drawing board, but as the refinement of
rough ideas that were first tried out in practice. The model was then
tested with static loads by the engineering department for strength and, if
necessary, adapted to meet Idflieg standards. Calculations were made
parallel to the practical testing to check the empirical data thus collected.
Formal design blueprints were subsequently made after the tested models. In so doing, Fokker; his designer, Martin Kreutzer; and, later, Franz
Möser deviated little from practices still common in aeronautical construction of the day, in which an empirical approach ruled over theoretical understanding. In all, the Fokker team would construct some 110 different prototypes during the fifty-one months the war raged, an average
of about one new prototype every two weeks.[17] This extremely empirical

approach, combined with the fact that the Fokker types of 1914 closely followed the Morane-Saulnier designs, brought Tony Fokker a remarkable commission from the German army, which would shape his future as one of the leading constructors in the German aviation industry.

## SYNCHRONIZING THE GUN

In the early months of 1915, the war on the Western Front moved into the air in earnest. France increased its aircraft production by about 60 percent between January and April, making for some 390 operational aircraft by the end of March. An additional eight-five and ninety were flown by the British Royal Flying Corps. Moreover, the French rearranged the deployment of their aerial forces, forming specialized pursuit squadrons, whose task it was to chase the enemy machines from the skies over the Allied trenches. They were equipped with such aircraft as the light, maneuverable Morane-Saulnier single-seater. As a result, the Allies could claim air superiority in the spring of 1915.

During the preceding winter months, the former Morane-Saulnier test pilot, Roland Garros, and his mechanic, Jules Hue, had been conducting experiments on Garros's plane, fitting it with a fixed machine gun that fired forward through the propeller arc. In doing so, they followed aircraft constructor Raymond Saulnier's prewar attempts to develop a synchronizing mechanism. Garros and Hue discovered that only 10 percent of all bullets that were fired actually hit the airscrew. Perhaps the propeller could be reinforced to withstand the bullets that hit it, enabling the fixed machine gun to be of practical use in air combat in a single-seat aircraft. Garros and Hue fitted the propeller blades of their Morane-Saulnier with steel wedges that deflected the machine-gun bullets. Their method was crude and risky (the bullets were deflected in various directions), but it worked. In two and a half weeks of operational testing over the trenches near Dunkirk, between April 1 and 18, 1915, Garros managed to shoot down three German airplanes—an exceptional score for the time, which added to Garros's renown.[18]

On April 19, his luck ran out. Garros's machine was hit by ground fire when flying behind the German lines. Both the pilot and the airplane, with its special contraption, were captured. Predictably, the machine gun and the unusual attachments to Garros's propeller blades attracted immediate attention. Various parts of the aircraft were subsequently sent to the army aircraft testing facility at Döberitz to be copied and tested. If the French had really learned the secret of shooting through the propeller arc so that the pilot could aim his gun by simply steering his airplane, the possible consequences for the German air arm could be very grave indeed. In view of existing French air superiority over the western battlefields, adapting the system for use on German aircraft was given high priority. Yet in tests conducted at Döberitz, the Germans discovered that simply copying the Garros mechanism did not do the trick. German standard bullets were capped with chrome tips instead of the softer copper that the French used, and they shot the wedges and the propeller blades to pieces. It was clear that a different solution had to be found.

The question was, who to approach to resolve the issue? Several constructors were invited to Döberitz to investigate the French gun, among them, Tony Fokker. At first glance, inviting Fokker was somewhat surprising given his thus far marginal position in the German aero industry. Yet in distant Schwerin, Fokker had been experimenting with various approaches toward a method of firing a gun through the arc of a rotating propeller in late 1914—experiments of which Idflieg must have been aware.[19]

Fokker was by no means the first to tackle the synchronization problem, and various patents on such mechanisms had been taken out in France, Britain, Russia, and Germany before the war. In Germany the idea of installing a fixed, forward-firing gun on an airplane had first been patented in 1910 by August Euler. His patent was published in May 1912. A patent had also been taken out on the idea of synchronizing the gun with the propeller's revolutions. In July 1913 the Swiss aeronautical engineer Franz Schneider, employed by the Luft Verkehrs Gesellschaft (LVG) in Berlin, had applied for such a patent. The German Patent Office took a year to study the principle before allocating German patent number 276396 to Schneider on July 10, 1914.[20] Schneider's solution to the problem was to install a cam on the propeller shaft that activated a trigger-blocking device every time the propeller passed the muzzle. Yet with the engine turning the propeller at about 1,200 rpm, a two-bladed propeller needed to block the gun mechanism 2,400 times each minute, that is, forty times per second. Because this would, in effect, prevent the gun from being fired at all, the Schneider solution appeared impractical and was never tested. On the basis of information thus available in the industry, Tony Fokker, with a small team headed by Heinrich Lübbe (a fairly well-known aviator of prewar days who had crossed over to aeronautical engineering from his profession as a watchmaker), took the two German patents as starting points for their researches. By May 1915 the team had practically solved the problem, but they lacked a suitable machine gun to complete the project.

Fokker duly arrived at Döberitz on May 19. In his characteristically brazen way, he assured the army officers of success and was allowed to leave for Schwerin with the propeller of Garros's airplane, a newly developed lightweight Parabellum machine gun, and ammunition. They set to work immediately. What the Fokker team came up with was a variation on the Schneider patent: what if the device were used not to block the gun, but to activate the trigger whenever the propeller was not directly in front of the muzzle? To prevent hitting the propeller, the gun would fire only within a safe sector between the revolving propeller blades. The mechanism delayed firing the gun as the propeller blades passed the muzzle—in the simplest terms, pulling the trigger instead of blocking it. The gear operated by a system of push rods, *Stangensteuerung* in German, which translated the cam action to the gun.[21]

This was precisely the kind of practical engineering in which Fokker excelled. With the basic principles established, linking the Parabellum gun to the engine of one of the M.5's under construction and synchronizing the gun with the propeller's revolutions took only two days. A wooden disc with a blackened outline of the airscrew replaced the propeller so that the impact pattern of the gun could be traced. The Fokker team fired the machine gun repeatedly until they determined how to position the cams on the axle to interrupt firing and allow the propeller blades to pass undamaged. Fokker had left Döberitz on May 19. When the sun set two days later, he lashed the tailskid of his modified M.5K/MG to the back of his car and drove to Berlin to demonstrate the invention that would revolutionize air warfare.[22]

Convincing his skeptical audience of military officers that the problem had already been solved turned out to be more time-consuming than adapting the gun mechanism. Not only was Fokker asked to demonstrate the device in flight at Döberitz, he also had to travel to the headquarters of the German army at Laon, in northeastern France, to have the synchronization system tested under combat conditions. Eventually, he was told to demonstrate his invention personally before Crown Prince Wilhelm, commander of the German Fifth Army, on May 23, 1915, at his headquarters at Stenay, some thirty miles north of Verdun.

Wilhelm and his staff were impressed but not entirely convinced, and the army's demand to demonstrate the device under actual combat conditions stood. Tony Fokker was fitted out in the uniform of a lieutenant of the German Flying Corps, given the proper identification papers for his temporary rank, and sent to the front to tour some of the German units with his aircraft. Sending one of the country's industrialists to risk his life proving that an invention worked was an extremely odd thing to do, even in wartime conditions. Fokker had the good fortune not to encounter any enemy airplanes during his demonstration tour. Had he not had his photograph taken, the story would have been too fantastic to be credible.[23] Because Fokker wanted to see the war in action and the enemy remained unseen in the air, he asked and was granted permission to tour the German trenches one morning, accompanied by another officer from the air corps—a frightening experience:

> I found the situation so interesting that time slipped by quickly. It was 10:30 before anyone realized it. The commander urged me to go at once, saying he would send a soldier back as far as the motorcar, which had been left at the far side of a demolished village. On the way through the trenches we were delayed, partly because of my curiosity. Just before reaching the abandoned village the expected artillery fire opened up. The screaming of shells as they approached; the bursting roar as they hit; the blasting of earth, bricks, and trees simply frightened me to death. I began dodging with every detonation, even though my mind told me that once the explosion was heard, it was too late to dodge. With one accord we started running, catching our feet in stray bits of wire, falling down in the mud and getting up even more scared. My arms were covered with filth clear up to my shoulders where I had plunged headlong into the muck. In a few minutes we were dripping. . . . The soldier who had brought us out of the trenches . . . was as near to laughing as a German soldier could come to mocking a superior officer. It made me feel pretty small, but none the less afraid.[24]

Fokker and the other pilot fled for their lives, only to be told later that what they had gone through was just a light artillery barrage. "But it was heavy enough for me," Fokker said.[25] The attack cured him of his wish to experience the war at close quarters. After several more days of flying, he wisely steered away from a French observation aircraft when he encountered one. Instead, a German military pilot, Oswald Boelcke, was sent up to fire the synchronized gun in actual combat and managed to hit an enemy airplane on his third trial flight. With this result, the initial skepticism suddenly dissolved. Orders for the synchronizing gear were soon coming in, and demand for the apparatus surged beyond Fokker's limited production capacity at Schwerin. For a while the army considered equipping non-Fokker types with the device. With the basic problem solved, it was not surprising that competing arms manufacturers, encouraged by the War Ministry's policy of reducing patent protection, responded by developing and promoting systems of their own before long.[26]

In response, Fokker developed an improved, simplified system, driven directly by the engine, that could be partly standardized to replace the sixty-odd different types of gears that were being built for different gun/aircraft/engine combinations. With cooperation from the Idflieg authorities, he was thus able to reduce the number of different gearing mechanisms to two: one for in-line engines, and one for rotary engines.

Fokker was now confronted with another bottleneck. Because of an insufficient supply of lightweight Parabellum guns, various types of heavier Spandau guns had to be adapted for aircraft use with the Fokker firing system to meet demand. In late June 1916, Fokker responded to these manufacturing difficulties by separating aircraft and gun production. He took over the small F. H. Zimmermann GmbH armaments factory in Reinickendorf, one of Berlin's suburbs, and renamed it the Fokker Flugzeug-Waffen Fabrik. By the time production started there in December,[27] competition was already fierce. For its own fighter-types, Albatros had different synchronizers built elsewhere. Siemens, a very large engineering and electrical firm near Berlin that had good contacts with the army purchasing department, also sought to cut into Fokker's market and initiated production of its own electrical synchronizing gears, on which the company had taken out a basic patent prior to Fokker's invention.[28] Fokker's countermove was surprisingly original. Hearing of the preproduction problems his new competitor was suffering from, he temporarily slowed down his own production, thus creating an artificial shortage of gun mechanisms that would bring Siemens's problems, and the risk the army took in supporting the venture, into the limelight.

Though heavy pressure was brought to bear on Fokker to abort this odd maneuver there and then, his approach worked. As a result, Fokker found himself in a more advantageous negotiating position vis-à-vis the army and was able to secure a six-month contract at a better price. However, the new contract came with certain conditions. The authorities demanded that Fokker fully separate his aircraft and armaments manufacturing activities and relinquish direct control over the Fokker

Flugzeug-Waffen Fabrik. Heinrich Lübbe, who had become Fokker's armaments engineer and deputy director at Reinickendorf, was put to work on a further improved version of the synchronization mechanism that would be easier to maintain and adjust under combat conditions. The new device, called *Zentralsteuerung* (central control), consisted of a flexible drive shaft linked to a gearbox driven by the engine, instead of the earlier lever and cam of the *Stangensteuerung.* In the spring of 1918, series production of this latest model began.[29]

Maintaining his position in armaments manufacture was by no means Tony Fokker's only problem with the synchronizing gear. More dangerous still, because it threatened to stop the cash flow from the invention, was the fact that Franz Schneider, the inventor of the original gun-blocking system, filed a lawsuit against Fokker for infringement of his 1914 patent. Knowledge of the Fokker invention had spread rapidly through the industry, and by early 1916 it was apparent that the army would order large-scale production of the device. In July, after six months of legal threats, Schneider brought his case before a Berlin court. To make sure he was suing the legal holder of the rights to the Fokker invention, he filed two claims: one against Tony Fokker personally, and one against Fokker Flugzeugwerke GmbH. Fokker, absolutely convinced that his system constituted a genuinely new invention to which Schneider had no rights whatsoever, and intent on holding on to the flood of orders that followed the successes of the first prototype gun gears at the front, was determined to fight the Schneider claim all the way. In a countersuit, Fokker's three patent lawyers demanded the Schneider patent be declared null and void because it had not been demonstrated to work in actual practice. Fokker eventually lost this case in March 1918.[30]

Given Fokker's immediate resort to heavyweight patent lawyers, it must have been obvious to Schneider that there was no easy victory to be won. To prevent the costs of the lawsuit from bearing entirely on his own shoulders, he sold his patent rights to LVG on September 16, 1916. LVG went along with this arrangement because the lawsuit promised to generate substantial compensation with little risk. At the same time, it could also undermine Fokker's position as a competitor on the rise. After LVG's entry into the arena, the intensity of the legal proceedings escalated. Tony Fokker, though advised that his case was weak, continued to fight. With money coming in from the expanding production of gun gears, he could afford to have his lawyers work overtime and concentrate on stalling the court proceedings. Nevertheless, on June 30, 1917, the Berlin Königliches

Kammergericht (Royal Lawcourt) ruled infringement of Schneider's rights because of the similarity of the two systems, and Fokker was sentenced to pay LVG one million marks in damages. Refusing to give in, Fokker went higher up the legal ladder to have the verdict overruled, even though the chances of winning such an appeal were slim. Nevertheless, his lawyers were still successfully stalling the case when the war finally came to an end in November 1918.[31]

## AN INDUSTRY ON THE RISE

Synchronizing the aircraft machine gun had a dramatic impact, not just on Fokker's position as an arms manufacturer in Germany, but also on the war in the air. In July 1915 the first Fokker monoplanes armed with the synchronized machine gun arrived at the front. Equipped with the innovative weapons system, this Fokker E.I (E = *Eindecker*: monoplane) fighter, as the armed production version of the M.5K/MG was designated, gave the Germans an important qualitative superiority over the Allies. By December 1915 the Germans still had no more than about forty E.I fighters operational, but these somewhat underpowered and relatively slow airplanes enabled the German Flying Corps to gain the upper hand in the skies over the Western Front.[32]

Flown by Germany's best pilots, now rising above the anonymity of mass warfare to be immortalized as "aces," the E.I brought about the ominous "Fokker Scourge," in which Allied airmen were called "Fokker Fodder." Though often quoted by Fokker historians, these were exaggerated terms. Production at the Fokker plant in Schwerin averaged only twenty-one airplanes per month through December 1915, which restricted the impact of the Fokkers on the air war. Production peaked in August with thirty-six aircraft.[33] The first kill with a Fokker E-type was made on July 1, 1915. Well-known German aces such as Lt. Max Immelmann and Capt. Oswald Boelcke achieved their first victories in August.[34] Despite some unexplained accidents, which blemished the flying record of the E.I and pointed to problems with quality control in wing construction at the Fokker plant, Germany's top-scoring pilots, Immelmann and Boelcke, and later "Red Baron" Manfred von Richthofen, henceforth demanded to fly Fokker airplanes.[35]

For Fokker, this endorsement was an extremely important asset. An accomplished natural pilot himself, he was about the same age as

Germany's frontline airmen, spoke the same language, and was receptive
to their immediate demands regarding the performance of the aircraft
they flew. Because he lacked the kind of contacts in the procurement bureaucracy that secured a continuous flow of aircraft orders, his position
as a manufacturer depended heavily on personal contacts with, and recommendations of, Germany's celebrated crack fighter pilots.

The Allies, on their part, intensified efforts to develop an equivalent
synchronizing gear. With greater numbers of aircraft, France and Britain
were quick to regain air supremacy. Between the big offensive at Verdun
in February and the beginning of the battle on the Somme in July 1916,
the Fokker Scourge abated. Subsequent models of the E-series of fighters,
marred by the difficulties of German aero-engine production, lost ground
to new Allied biplane fighters, which appeared over the Western Front
early in 1916: aircraft such as the French Nieuport 11 and the Sopwith
1½-Strutter, which were faster and offered better climbing characteristics
and maneuverability than the Fokker fighters. On the advice of Boelcke
and other leading pilots, with whom Fokker had established a close
rapport during his "demonstration tour" in France, Fokker's designer—
Martin Kreutzer—and his team developed a biplane successor to the
E-series according to Idflieg specifications. Though the fame of Fokker's
early fighters had now placed the Fokker Flugzeugwerke at the forefront
of the German aircraft industry, the initial models of the new Fokker
D-series (D = *Doppeldecker:* biplane) of fighters proved disappointing.
Adding an extra wing meant adding weight, thus reducing engine efficiency. The Fokker D-series aircraft were outclassed by their more maneuverable Albatros and Halberstadt counterparts, which incorporated
such innovations as aileron controls, while the Fokker types continued to
depend on wing-warping. Consequently, Fokker lost out to Albatros and
Halberstadt in Idflieg's allocation of the much-coveted 120 and 160 hp
Mercedes aero engines. The Albatros D.I and D.II fighters became the
backbone of the German frontline units, representing 67 percent of their
equipment in January 1917. The quality control problems added to
Fokker's difficulties. His D.IV model, built as a trainer only, was even
grounded for poor workmanship and material. Problems with quality
control, usually surfacing as wing failures, were a recurrent phenomenon
in the German wartime aircraft industry. The otherwise first-rate Albatros
D.III and D.V of 1917 also suffered a series of such accidents.[36] In his autobiography, Fokker later blamed Idflieg for his temporary demise. The
agency failed to furnish him with the best available materials and engines,

and noted that problems with supply seemed to demand more and more of his time as the war dragged on.[37]

Judging engine development to be the key to recapturing some of the ground lost, and to spread his business risks at the same time, Fokker took a controlling interest in the Oberursel Motoren Werke AG in the summer of 1916 for a reported four million marks.[38] With the acquisition of the Reinickendorf plant and his stake in Oberursel, Tony Fokker, now twenty-six years old, was well under way to becoming a major industrialist, and his interests expanded rapidly. That summer he also acquired a substantial shareholding in the Ungarische Allgemeine Maschinenfabrik (Magyar Altalános Gepgyár RT [MAG]) in Budapest. By that time, he had a history of deliveries to the Austro-Hungarian Luftfahrttruppe. Between August 1914, when the Austrians had first approached him about the sale of fighter aircraft, and the time the Luftfahrttruppe ceased purchasing aircraft in Germany in mid-1916, Fokker delivered 108 airplanes to the Dual Monarchy. After the Austro-Hungarian change in procurement policy, Fokker was quick to invest in MAG, which had established an aircraft production department some months before the policy shift. The same month, on August 26, MAG received a contract for 50 Fokker fighters from the Luftfahrttruppe. The first of these aircraft was accepted in March 1917.[39] Yet, curiously enough after acquiring a stake in MAG, Fokker appeared to care little about the fortunes of his Hungarian venture and the possibilities of boosting deliveries to the Austro-Hungarian air corps. As with Fokker's other investments, he had staked money on MAG primarily to diversify his business interests. His aircraft production manager at MAG, Friedrich Seekatz, who had joined the Fokker team in August 1914 as a supervising engineer and had been posted to Budapest from Schwerin, complained bitterly about this lack of interest and the insufficient means to boost production.[40]

In his various business ventures, Fokker followed a somewhat peculiar policy: rather than investing profits in his own company, he preferred to buy stakes in businesses related to the Fokker works. His investments were those of a reluctant industrialist, suggesting that he was losing his grip on the key goal he had set out to achieve: building Fokker airplanes. Shunning further investments in the Fokker Flugzeugwerke, Fokker allowed his Schwerin company, now in the capable hands of his new business manager, Wilhelm Horter, to struggle with the problems of wartime aircraft production and development. Meeting increased demand and the productivity drives that ensued from the total mobilization of Germany's

industrial apparatus in the Hindenburg Program of October 1916 by increasing his work force, Fokker preferred to put up new makeshift sheds
rather than invest in permanent factory expansion. Likewise, he acquired
the nearby Perzina Pianoforte Fabrik and the Pianoforte Fabrik Nütz-
mann for just under two million marks and diverted part of his wing production to their workshops.[41] By October 1917, when a complete training course in most aspects of aircraft construction still took no more than
eight weeks, the Fokker Werke employed over one thousand people.[42]

Nevertheless, 1916 was a difficult year. As the performance, and hence
production figures, of his various D-series biplane fighters remained unsatisfactory, the Fokker Werke itself went through something of a crisis.[43]
New fighter types, notably the D.I and D.IV biplanes, remained under a
cloud of structural and engineering problems and consequently failed to
attract orders until Idflieg criticisms of quality control had been satisfied.
To make matters worse, chief designer Martin Kreutzer died in a crash in
a Fokker D.I on June 27, leaving a hard-to-fill void at the heart of the
company. While Reinhold Platz, Fokker's long-standing hands-on companion, remained in charge of prototype construction, design responsibilities came to be shared by Fokker; his new chief designer, Franz Möser,
and the engineering department; and occasional outside experts such as
the Swedish engineer Villehad Forssman.

Late in March 1916, Fokker, on the lookout for new ideas to bring his
aircraft back into the limelight, had met Forssman in Berlin, where they
had discussed the latter's ideas about the possibility of using wings covered with plywood veneer. Forssman pointed out that such wings offered
important advantages over those made by the traditional construction
method, consisting of a wooden framework covered with fabric, because
they would have a higher torsion stiffness and be lighter, as well. Seizing
on this unexpected possibility, Fokker and Forssman closed a secret deal
under which Forssman was to supply a "veneer wing" for Fokker's M.20
prototype. In the next five months, Forssman built three sets of plywood-
covered wings for three different Fokker designs, but despite such innovations the road to series production of Fokker fighters embodying the
new wing design proved a long one. Not until the end of 1917 did
Fokker incorporate the new wing design in series aircraft.[44]

By January 1917 Fokker's position in German aircraft construction
had eroded to the extent that he was instructed to build 200 AEG C.IV
training aircraft instead of producing new Fokker fighters. In June a second contract followed for another 200 such machines. The first of the

AEG airplanes was delivered on August 1, 1917. Possibly because of his changing position in German fighter aircraft production, which was strongly dominated by Albatros and LVG, and because he feared that a German defeat in the war would bring a collapse of the German mark and make his fortune disappear, Fokker further diversified his business interests and took up a shareholding in the Axial propeller factory. As a private individual he also invested in a small seaplane factory on the Baltic coast: Flugzeugwerft Lübeck-Travemünde. To supervise these investments, he founded the Fokker Zentralbüro (Central Administrative Office) in the Hotel Bristol in Berlin, his customary foothold in the German capital.[45]

The reduced importance of Fokker designs was evident in a surplus stock of his D.III fighters, which Idflieg had chosen not to dispatch to frontline units. Informal cross-border contacts with the Dutch, who wanted to supply the meager Netherlands Air Corps with more up-to-date equipment, led to the arrival of a delegation in Berlin early in August 1917 to view the then already obsolete aircraft. Dutch neutrality in the war meant that the negotiations had to be kept low-key, but nevertheless, an agreement was reached with Idflieg whereby Fokker was permitted to export ten (initially twelve) D.III fighters to Holland. The first D.III arrived on October 1, 1917.[46] By the end of the year, the deteriorating military position of Germany in the west formed the backdrop for talks on an export license for an additional six D.III's, along with extra Oberursel engines and spare parts. This issue was linked to German attempts to buy some five thousand horses from Holland that were badly needed by the army in exchange for forty Rumpler C.VIII training aircraft to be delivered in 1918.[47] Because the export of such a large number of horses would have constituted a serious breach of Holland's neutrality and might even have led to Dutch involvement in the war on the German side, nothing came of the exchange. The forty C.VIII's were eventually delivered for cash: one batch of eight aircraft in 1918, the other thirty-two machines in 1919. Otherwise, the Dutch War Office rejected offers to buy German military aircraft.[48]

Also in 1917, Fokker entered into a joint venture with Professor Hugo Junkers. The initiative for this move had come from Maj. Felix Wagenführ of Idflieg. Wagenführ, responsible for aircraft planning, was looking for a production facility to build a revolutionary all-metal armored ground-attack aircraft that Junkers was developing at that time. Junkers had already completed two prototypes, the J.I and the J.II, of these cantilever-winged monoplanes. The aircraft were intended for use as close-support fighters over the trenches at the front. With Fokker fallen from grace as a producer, his factory was just what Idflieg was looking for in the context of the America Program, anticipating the adverse effects of entry into the war by the United States.[49] On December 16, 1916, Wagenführ had called for a meeting between parties in Berlin.

With his usual keen sense for business, Tony Fokker immediately announced he was very interested in the Junkers designs but unable to start producing Junkers aircraft in his factory, as Idflieg had in mind. Instead, Fokker proposed that he and Junkers combine forces to bring the prototypes to perfection and ensure an optimum setup for mass production of the aircraft. Two days later, he visited the Junkers plant in Dessau, seventy miles south of Berlin, to inspect the aircraft. An enthusiastic Fokker was shown all plans and drawings of the new aircraft and given the opportunity to take the J.II on a test flight on December 22. Fokker then proposed to buy the rights for license production of the aircraft from Junkers for 500,000 marks and compensate Hugo Junkers personally for the use of the patents and the necessary cooperation in readying the J.II for series production.[50] However, Fokker did not agree to the sum that Junkers demanded, and in January 1917 he gradually started backtracking from the prospective deal. Behind this seemingly whimsical behavior lay a bold plan, for, when in Dessau, Fokker had received a full explanation of Junkers's revolutionary cantilever wing construction. Junkers and his proud team of engineers had told Fokker as much as he could grasp about this new development in aeronautical engineering, and rather than cooperating in the series production of Junkers aircraft, Fokker wished to exploit his newly acquired knowledge for designs of his own. Learning about Fokker's backtracking, Hugo Junkers was furious, questioning Tony Fokker's personal integrity in no uncertain terms. Yet the damage had been done, and through his business manager, Wilhelm Horter, Fokker, who now steered as far clear from Junkers as he could, let it be known that on second thought he only wished to acquire the right to use Junkers's patented cantilever wing design.[51]

Much of what followed in the subsequent months was damage control. Under Idflieg's directions, Junkers and Fokker negotiated an agreement that linked the two men and their companies. On June 16, 1917, a contract was signed under which Fokker acquired the rights to license production and further development of aircraft after Junkers's basic designs, including the use of the cantilever wing construction. Fokker agreed to pay 500,000 marks for this, plus a personal commission to Hugo Junkers of 10 percent of the price of each airframe thus built in the Fokker Works for the duration of the war (and 7.5 percent after the war, until February 1, 1925).[52]

In August the two men also reached agreement on founding a joint company, Junkers-Fokker Werke AG Metallflugzeugbau. The idea was that Junkers's experience in theoretical aerodynamics and innovative all-metal cantilever-winged aircraft would join Tony Fokker's mass production know-how, pragmatism in design, and exceptional skills as a test pilot. Idflieg hoped that the linkup would be instrumental in persuading Junkers to simplify his designs in the interest of increased production at his Dessau plant. The new venture was founded in Dessau on October 20, 1917, after further months of preparatory groundwork. Total capitalization stood at 2,630,000 marks, of which 630,000 came in the form of a subsidy from the army. Junkers provided all real estate, fixed assets, factory equipment, and stocks from his existing Dessau plant. The shareholding of the company amounted to two million marks, divided equally between Hugo Junkers and Tony Fokker. The latter agreed to buy his stock from Junkers at a course of 118.5 percent. Fokker carried the title of director and was charged with boosting production, while Junkers figured as designer. The Junkers-Fokker Werke was the exclusive recipient of the rights to use existing and future Junkers patents for the production of metal aircraft of Junkers design, though further license rights could be allocated to the Fokker Flugzeugwerke in Schwerin and Junkers's parent company, Junkers & Co., for a nominal fee of 250,000 marks plus 9 percent of the net price per aircraft built.[53]

Something of a shotgun marriage, the Junkers-Fokker partnership was characterized by misunderstandings. Due to the novelty of the all-metal construction and the complexity of the Junkers design, the company had severe difficulties meeting promised delivery dates, and Idflieg repeatedly threatened to cancel price agreements for the aircraft.[54] Nevertheless, 227 Junkers J.I armored ground-attack aircraft were delivered in 1917 and 1918, plus 41 Junkers D.I all-metal fighters around the end of the war. The company also produced artillery fuses, field kitchens, and sterilization units.[55]

Fokker's main interest, however, was the use of the Junkers patent on the cantilever wing construction. Cross-fertilization of ideas went beyond the conceptual resemblance of Fokker's V.17, V.20, V.23, and V.25 cantilever-winged monoplane fighter prototypes of early 1918 and

Junkers's similar design approaches in the J.7 (flown and crash-landed by Fokker on December 4, 1917) and J.9 of the same period.[56] Though Fokker used wood for his wing construction instead of the aluminum preferred by Junkers, the similarity of the two concepts was evident in legal proceedings between Junkers and Fokker over the patent rights to the cantilever wing construction that dragged on until 1940.[57] In light of these developments, it was not surprising that the joint venture ended in bad blood and was dissolved by mutual agreement on April 24, 1919. According to Fokker's estimate, the venture cost him a total of 1.5 million marks. Junkers took over all assets and continued to operate under the name of Junkers Flugzeugwerke AG.[58]

By the time Junkers-Fokker Werke was dissolved, Fokker had long since combined two important novel approaches to aircraft design that he came across in 1916 and which characterized Fokker aircraft from 1917 onward: Forssman's veneer-plywood-covered wings, and Junkers's cantilever construction. Late in 1916, Fokker and Möser designed their own cantilever wing constructed from wood. It was at once lightweight, efficient, *and* strong. Proceeding from ideas and sketches to an actual wing took place in the usual trial-and-error method with which Fokker liked to approach questions of technological development.[59] Linking this new wing to an experimental, highly streamlined fuselage produced a revolutionary cantilever wooden-winged biplane identified as V.1 (V = *Versuchsmaschine:* trial machine). It was a breakthrough design of the first magnitude. Unfortunately, given the production-capacity and horsepower-output problems that German aero-engine manufacturing suffered from throughout the war, a suitably powerful engine could not be provided for the V.1. This, combined with poor visibility from the cockpit, made the aircraft unsuited for military use. Even though no production followed, the general concept of the cantilever wing had been proven to work. Unchanged in its basic design and structural characteristics, it became the trademark of all subsequent Fokker designs for the next two decades.

What saved Fokker as a major producer of fighter aircraft was the appearance of the Sopwith Triplane over the Western Front late in January 1917. In aerial dogfights, two factors were of overriding importance: altitude and speed. To climb faster and higher than the opposition and then attack, according to Boelcke's dictum, with the sun to the rear, were vital to the element of surprise, which was paramount in achieving success. Most aerial dogfights lasted only a few seconds, with the objective of placing as many bullets as possible in the enemy airplane before breaking away. Operational ceiling and the ability to achieve it quickly were, in those days, limited by two factors: engine effectiveness and the lift efficiency of the wings. In respect to the former, German frontline aircraft remained severely handicapped throughout the war because of the inability of the aero-engine industry to produce engines comparable to those built and used on the Allied side. In respect to the latter, adding an extra wing meant more lift, but at the expense of greater weight and reduced speed. The arrival of the British triplane caused frontline commanders such as Manfred von Richthofen to write anxious letters to Berlin demanding early reequipment with fighters that could match the perceived (though actually inflated) qualities of their opponents'. On July 27, 1917, Idflieg heeded these pressures and sent out an urgent circular to all aircraft manufacturers, asking for promising triplane fighter projects.[60]

By that time, Fokker was already in a good position to make a comeback in the upper levels of German aircraft construction. His fortunes had always relied on his contacts with pilots at the front and their willingness to lobby in Berlin for Fokker aircraft. Berlin's policy was to put an end to that influence, which presented the already overburdened procurement bureau with a wildcard that endangered systematic planning. Fighting for his position in the procurement circus, Fokker had set out for the Western Front on a fact-finding and goodwill mission after the first news of the British triplane reached him in April. Making the most of his personal acquaintance with Von Richthofen, he stayed several weeks at the ace's Jagdstaffel 11 (Fighter Squadron 11) in Courtrai, southwestern Belgium—long enough to be shown one of the first of the new British airplanes that had been brought down behind the German lines.[61]

He went back to Schwerin and began planning for a German triplane as an answer to the Sopwith. The biggest problem would be securing an engine that matched the power plant of the British machine, especially at higher altitudes, where the low-compression German engines lost out to the high-compression engines used by the French and the British. The only German engine that met his demands was the recently developed 110 hp Oberursel Ur.II, an adaptation of the French Le Rhône rotary engine. The Ur.II had only passed its type test in March 1917, and production examples of the new engine were few. Yet the need for the new high-performance rotary engine was so great that Idflieg decided to take all Le Rhône engines that could be salvaged from crashed or captured Allied airplanes and send them to a special workshop for repair and subsequent reuse. Repaired Le Rhône engines powered about 16 percent (fifty-one aircraft) of the subsequently produced Fokker triplanes.[62]

By early August, the Fokker triplane had evolved through several prototypes into a machine that could be tested as the Fokker F.I, later redesignated Dr.I (Dr = *Dreidecker:* triplane) by Idflieg at Adlershof/Johannisthal. A preliminary production order for twenty aircraft had already been placed by Idflieg on July 14.[63] The Dr.I passed the initial acceptance tests without serious structural problems. A pleased Tony Fokker personally demonstrated his new airplane before Quartermaster General Erich von Ludendorff, the highest weapons procurement authority, in Courtrai on August 26, 1917. Von Ludendorff was suitably impressed. Von Richthofen also evaluated the Dr.I and requested that his Jagdstaffel be reequipped with the new Fokker as soon as possible.

With such resoundingly distinguished names backing the little triplane, Fokker was offered an initial production contract for 320 aircraft, the first 173 to be rushed to the front no later than December 1, 1917. With these new orders, the Fokker Works suddenly entered an entirely new style of aircraft production. The purchase far exceeded Fokker's production capacity in Schwerin, burdened as it was by the AEG trainer order. An improvised series production was nonetheless set up, and by the middle of October the first seventeen Dr.I's were delivered to the Flying Corps squadrons operating over the Western Front.

Things did not continue to go smoothly for long. Within weeks several of the new aircraft crashed. Eyewitnesses reported that the top wing had come apart. The disasters recalled accidents with Fokker's E-types and D-types in the past, and a special investigative committee was rushed to the front to determine the cause of the crashes. It was discovered that aileron operation under combat conditions put excessive stress on the triplane's top-wing structure. The problem had been aggravated during production of the series aircraft, when, in great haste, Fokker neglected to apply protective varnish to the wooden internal structures of the wing. Frontline use of the Dr.I resulted in a buildup of moisture inside the wings, which affected the bonding of the glue. As a result, all the new Dr.I's had to be grounded temporarily, and during November the Fokker Flugzeugwerke replaced all previously delivered wings with new ones, setting back production substantially.[64] After these problems had been remedied in December 1917, the Dr.I achieved fame and immortality with Manfred von Richthofen, the Red Baron, who had his personal airplane painted a defiant red. Because follow-up orders had been canceled after the crashes, production totaled 320 aircraft until ending early in June 1918. Attempts to develop an improved-performance Dr.I, using newly developed high-powered rotary engines, failed to produce a viable replacement.

## THE ULTIMATE FIGHTER AIRCRAFT

By the summer of 1918, Fokker already had a new fighter in production. Late in 1917, his engineering team had endeavored to apply the cantilever wing construction to a biplane configuration. Again, one of the biggest problems was where to procure the high-performance engines necessary for the airplane to match the latest versions of the French Nieuport and Spad fighters, the British Royal Aircraft Factory SE.5A, and the Sopwith Camel. Several fighter prototypes were developed simultaneously, and of the two basic configurations, one was powered by rotary engines, the other by the more powerful watercooled in-line engines. In January 1918 these prototypes were nearing completion, along with several other aircraft projects. In response to a petition from Germany's top frontline pilots to select its next standardized fighter type, Idflieg held a special fighter competition at Adlershof on February 19, 1918. Fokker entered eight different aircraft: four biplanes, three triplanes, and one monoplane. Much was at stake, and like his competitors, Pfaltz and Albatros, Fokker did his best to bend the rules a little in his favor by entertaining military pilots "in style" with "gaiety, charm, diversion, the society of pretty girls, the kind of a good time they had been dreaming about during their nightmare stay at the Front. Berlin was full of girls eager to provide this companionship."[65]

Besides, Fokker had learned how to oil the hinges of bureaucracy. After the first flights with one of his biplanes, he spent a somewhat secretive weekend with several faithful compatriots changing the construction details of one of his entries in the Adlershof competition. Continued support from Fokker's personal friend, Von Richthofen, and other crack pilots on the front also worked in Fokker's favor. Of his two cantilever biplane designs, one with a rotary and one with an in-line engine, the latter, known as V.11, won the 160 hp Mercedes-engine category.[66] The champion aircraft was redesignated D.VII.

This success immediately put the next problem before Fokker: how to build large numbers of the badly needed fighters quickly. Idflieg, well aware of the limits of Fokker's production capacity, decided to play it safe and, in an unprecedented step, ordered the D.VII to be built in large numbers by Fokker's main competitor, the Albatros Werke, and its subsidiary, the Ostdeutsche Albatros Werke. Fokker was to receive 5 percent of the airframe price for every license-built aircraft. Of the initial order of no fewer than 700 D.VII's in February 1918, 400 were to be built by Albatros, against only 300 by Fokker in Schwerin. Fokker's remaining AEG C.IV production was canceled. Though this solution enabled Idflieg to circumvent Fokker's production capacity, it created its own problems. Licensed production of the D.VII was hampered by deviations in construction practice. As a consequence, D.VII's built at the different works had only a limited number of interchangeable components.

Despite continued complaints about quality control in Schwerin, Fokker had now arrived at the pinnacle of the German aircraft manufacturing industry.[67] On May 1, 1918, nineteen D.VII's were operational over the Western Front; four months later, there were 828. Licensed production of the D.VII had made this possible.[68]

Meanwhile, Fokker's latest venture in fighter aircraft, the E.V (also known as the D.VIII), dubbed "The Flying Razor," once again proved controversial because of structural problems. In August 1918 several of the newly delivered lightweight fighters crashed. As before, the difficulties appeared to lie in the construction of the wings. Because the E.V was the first Fokker type to incorporate Forssman's glued-on sheets of plywood for wing covers, its wings attracted attention to begin with. And as before, it was found that poor workmanship, defective materials, and general lack of effective quality control were to blame. Water accumulated under the plywood surfaces, affecting the bonding capacity of the glue that held the wing structure together. One Idflieg report in August 1918 spoke of "thick streams of water" that ran out of the wings after holes had been bored in them.[69] Once again, a Fokker fighter was grounded. Fokker, accused of at least facilitating if not actually invoking constructional errors resulting from shoddy production practices, was forced to modify the wings and ensure proper supervision of production and quality control. The E.V was withdrawn from the front after only three or four days of operational deployment in August and did not return before the end of hostilities.[70]

Thus Fokker's D.VII remained in service as Germany's ultimate fighter airplane. Despite its rather angular and unpolished looks, the aircraft had excellent flying characteristics and caused quite a stir among Allied airmen. It soon ranked as the most-feared of the German fighters, so much so that the Fokker D.VII was specifically mentioned in the Armistice Agreement of November 11, 1918. Article 4 read:

> Abandonment by the German Armed Forces of the following war materials in proper condition:
>     5,000 cannons (2,500 heavy guns and 2,500 field guns),
>     25,000 machine guns,
>     3,000 mine throwers,
>     1,700 fighter and bomber aircraft, *beginning with all D.VII aircraft* and all night-bombing aircraft, to be turned over on the spot to the forces of the Allies and the United States—under the detailed conditions stipulated in the Annex n° I[1], to be seized at the moment of the signing of the Armistice Agreement.[71]

This uninvited attention and praise spelled trouble in the making. Tony Fokker, the former dropout, had made a name for himself that reached far beyond the aviation arena. Alternately feared and admired, the Fokker name had become a household word, symbolizing a fearsome fighting machine. The armistice, and the peace conditions that were subsequently negotiated in Versailles, left no doubt that the Allies wished to erase Germany as an aeronautical power. Its aviation industry was to be dismantled. As Fokker watched his market collapse after the armistice, his career as an aeronautical constructor in Germany was in jeopardy. In no mood to be forced to retire from aviation, he secretly began to think about moving part of his factory from Schwerin to Holland. There, Fokker had high hopes of being welcomed, especially because of the virtual absence of a Dutch aircraft industry, and because the Dutch Air Corps was very pleased with the few Fokker warplanes they had been allowed to buy in 1917.

# 3

## CORNERED AMID CHAOS

Throughout the four years of constant turmoil during the war, Tony Fokker's lifestyle was unsettled. Though he came to own several pieces of real estate, he remained a lodger on the first floor of a rooming house in Schwerin run by an elderly lady, Mrs. Frieda Grabitz. There he and his friend Bernard de Waal (in charge of the Fokker flying school) lived a not uncomfortable, if somewhat Spartan, life. "For me money is nothing else but a tool, indispensable for development in the way that I want. I spend little on myself, only as much as is necessary to feel pleasant, without throwing it away on follies or luxury the way hundreds of war suppliers do in Berlin. . . . Nevertheless, I have got everything I need for the relaxation of body and soul.[1]

Fokker's peculiar housing situation reflected some uncertainty about what his future might bring. Playing it safe was a constant in his behavior. In August 1914 he had surprised his associate Friedrich Seekatz by buying six pairs of shoes at once, stating that one never knew how long the war would last.[2]

In response to the many pressures and uncertainties of wartime production, Tony Fokker developed his own, uncompromising lifestyle. He was something of a workaholic. With no family to attend to, he worked all hours of the day and the night if he felt like it, or needed to. For meetings with his management staff at the Fokker Zentralbüro in Berlin, or with ministry officials, he customarily stayed at the Hotel Bristol on

Berlin's most prominent avenue, Unter den Linden, where he occupied three second-floor rooms on a permanent basis.[3] As a young, nonsmoking, nondrinking, and eligible bachelor, he led a curious life of abstinence from the pleasures that were there for the taking. He preferred the camaraderie of frontline pilots who visited Berlin or Schwerin when on leave. Oswald Boelcke, another teetotaler, who shared Fokker's fascination with aviation and practical technology, was one of his closer friends. His death in action on October 28, 1916, left a void that went beyond the loss of a sympathetic supporter of Fokker's products: it was the loss of one of his few real friends. Although not absent from his life, women remained enigmas he would be infatuated with time and time again, even though no serious commitments ensued. In October 1917 Fokker wrote to his father, "I have no time for women yet, and a woman would get little out of me, as I completely submerge myself in my work, that consumes all of my time nowadays, and the development of which is my life's goal."[4]

In spite of this avowed dedication to his work, in the spring of 1916 Fokker had named his sailing yacht *Tetta*[5] after his favorite girlfriend, Tetta von Morgen, the daughter of Lt. Gen. Ernst Curt von Morgen, who commanded an army unit on the Western Front, and his deceased wife, Maria Guthmann. They had met on the Wannsee, a lake just outside Berlin and a popular boating resort for Berlin's society. Instead of formally introducing himself to her, Fokker admired Sophie Marie Elisabeth von Morgen (her full name) from a distance until, one day, fate would have it that he was nearby when she lost her balance and fell overboard from her yacht. Jumping to her "rescue"—she was a good swimmer, Fokker admitted—gave him the introduction he was waiting for.[6] After this dramatic meeting, the relationship between Tony and the athletic twenty-year-old brunette, born in Zehlendorf, near Wannsee, on August 12, 1895, developed along rather more sedate patterns, and its progress depended on the frequency of his visits to the Wannsee. A formal engagement to be married did not follow until the spring of 1918.

In the meantime, Fokker's loyalties wavered. In *The Wind and Beyond,* aeronautical engineer Theodore von Kármán, one of the board members of MAG, remembered,

> He was . . . constantly embroiled in love affairs. I remember that once we travelled by train from Vienna to Budapest for an important conference. When the train arrived at the station in Budapest, I looked around for Fokker. He was nowhere to be seen.
>
> "Where is Herr Fokker?" I asked his assistant anxiously.
>
> "Oh," he said, "Herr Fokker has left the train."

"But I told him it was an important meeting of the Board of Directors. What happened?"

The assistant smiled. "Herr Fokker met a pretty girl and left with her. He anticipated you might get angry," he added, "so he told me to tell you that a meeting of the Board of Directors will take place every month. But if you let a beautiful young lady slip through your fingers, you will never get her back again."[7]

Nevertheless, the young lady, like numerous others, soon slipped through his fingers all the same. Through a consequence of a number of unlucky brushes with members of the opposite sex, Tony Fokker had only his little dog, Zeiten, a black, long-haired dachshund, as a constant companion. In his autobiography he remembered Zeiten fondly, calling him "the most intelligent dog I've ever seen."[8]

Boating, besides offering an opportunity to see Tetta from time to time, was an informal backdrop for business deals. Even more important: then and in his later life, the water was Fokker's preferred environment for relaxation. In July 1917 he passed the ballot and was greeted as a full member of the Wannsee Sailing Club, Verein Seglerhaus am Wannssee. Fokker was an eminent sailor and a popular member, repeatedly asked to enter competitions. For a man of his means, sailing was a relatively cheap pastime, befitting his reluctance to spend money on personal luxury. The annual club fee for 1918 was 120 marks. The *real* expense lay in the ships themselves, and here Fokker indulged in the luxuries that money could buy. In addition to the *Tetta*, he owned several smaller racing yachts that he named after himself, like the twenty-three-foot *Fok. III* (formerly called *Melusine*), which he bought in June 1917, when prices for such nonessentials were falling because of the general deterioration of economic conditions in Germany. *Fok. III* was followed in October 1917 by a racing yacht of similar size that he rechristened *Fok. IV*, and by the *Fok. V*, which he acquired in the late spring of 1918.[9]

Characteristically, he did not obtain the *Stander Zertifikat* for these smaller vessels, which was required to carry the triangular red and black banner of the sailing club. The club's secretary had to remind Fokker repeatedly of this obligation as late as June 1918, stressing that he needed to put his papers (official measurements and classification) in order if he wanted to enter the ships in any contest.[10] With German fortunes in the war declining, and shortages that increasingly contributed to a dramatic inflation of the German mark, Fokker started putting more and more of his earnings into low-risk personal properties of constant value. During

1918 he bought a large house near the Berlin Zoo, for which he paid one million marks, lots of antique furniture, a country house ten miles outside Berlin, and another in southern Bavaria.[11] And he bought yet another boat, this time a 6-meter yacht he renamed *Puschka* (originally *Schelm*), reminiscent of unforgotten love for Ljuba Galantschikova. Like his other boats, the *Puschka* was moored at the Verein Seglerhaus am Wannsee. With it Fokker participated very successfully in the annual south German regatta at the Starnberger See, which was held as if no war existed, between July 20 and 28, 1918.[12]

Diversions aside, signs that things were seriously awry in Germany were now unmistakable. The German-Soviet Peace Treaty of Brest-Litovsk (March 1918) was too late to have the decisive effect on the economy, morale, and troop strength needed for the success of Germany's ultimate spring offensive in the west. The prospect of defeat loomed ever larger. In the summer of 1918, the military tables turned irreversibly. By September 1918, defeat was clearly imminent, and the advance of the Allied armies was unlikely to be stopped.[13] Weary of war, Germany tumbled into crisis. After four years of rationing and substitute products, living conditions had deteriorated dramatically; stocking up on food and heating fuel for the oncoming winter preoccupied the minds of the civilian population. The German government, now gaining in authority after a power struggle with the military that had a history as long as the war itself, was preparing a peace offer.

Still, Berlin's society tried to maintain a "business as usual" attitude, and the Seglerhaus am Wannsee held its annual competition for members on two consecutive September Sundays. Fokker, just back from one of his frequent visits to his Budapest subsidiary, participated. The country was on the verge of collapse, yet those taking part seemed more worried about finding enough crew members to schedule the event on a single day. Not a month later, even their world was shaken when the last cash reserves of the sailing club ran out. Tony Fokker agreed to buy the club's large cruising yacht, *Erika*, to keep the Seglerhaus going.[14] Not for long: in view of the general shortage of food, the Reichskommissar für Fischversorgung (State Commissioner for Fish Supplies) issued a decree that the club's sailing boats had to be made available for fishing. Shipowners who did not comply ran the risk of having their ships seized. Like the other members, Fokker was asked which of his ships he "volunteered" to offer to the fishing program.[15] Preoccupied with more important things as his empire crumbled, Fokker didn't even bother to reply.

## ESCAPE TO HOLLAND

Ever since the defeat in the battle of Verdun in December 1916, German fortunes in the war on the Western Front had declined, a process that even a peace settlement with the Soviets could not counter. The big German offensive of 1918 brought only short-term tactical gains, which the army was unable to consolidate. In Berlin, the power struggle between the military and the civilian government deepened, reaching crisis level by October 4, 1918, when the cabinet of Prince Max von Baden made a armistice offer to United States president Woodrow Wilson in response to Wilson's famous Fourteen Points peace plan of January.[16] During the following weeks, common support for the peace initiative started taking on the shape of an active antiwar movement that was carried by the political left. The situation escalated when the High Command of the Imperial Navy decided to take to sea and engage the Allied forces in a decisive final battle. Dissatisfied seamen, unwilling to die for a cause already lost, started a mutiny in Kiel on November 4. The revolt, quickly focusing on peace and a more democratic form of government, spread rapidly, though the movement itself remained curiously local in character. Between November 6 and 9, councils of soldiers and workers sprang up overnight in cities throughout Germany and seized control, filling the vacuum brought about by the collapsing military and the power struggle in Berlin.

Most of the councils had no clear-cut objectives of achieving national dominion, as in the Soviet paragon. Their goals were often not even radical or left-wing, but a forceful protest against the breakup of worker solidarity caused by the war, and against the deep rifts between those in positions of authority and those on the lower steps of the social ladder that were so characteristic of German society. In response, the Reichstag tried to channel the general dissatisfaction into a reform of the system of government away from the traditional alliance between the military and society's elite, which had so willingly taken Germany into the war. Both the reform process and the escalating dissatisfaction in the streets were boosted by the ongoing military collapse, which destroyed the prestige (and thus the influence) of the army and its supporters. On November 7, 1918, the Social Democrats finally asked Emperor Wilhelm II to abdicate in order to facilitate the changeover to truly democratic, parliamentary government and thus escape from the potentially explosive situation.

As in many other cities in Germany during that first week of November 1918, Schwerin was taken over by a number of spontaneous workers' soviets. On the morning of November 6, several companies of the 89th Infantry Replacement Battalion in Schwerin marched to the Fokker factory in Görries and ordered the workers to strike. As elsewhere, a joint workers and soldiers council was set up. Besides recognition by the city of Schwerin and the removal of the old guard of high administrators, the council demanded acknowledgment of its self-appointed power as a police force and of its authority over the battalion's officer corps. Political prisoners were to be released immediately.[17] Tony Fokker, who had generally been uninterested in politics and not clearly aware of the significance of what was happening around him, was first shocked, next angry, and then terrified to suddenly find himself in the middle of a revolution. His authority and position as director of the Fokker Works was challenged. The workers' soviet formulated a series of immediate demands reflecting the months of labor unrest that had swept through Germany since April. Yet from Tony Fokker's perspective, its leaders were a bunch of "irresponsible dictators. For a time, we could not dismiss workmen without the consent of local authorities. . . . Things went from bad to worse. We feared that we would not escape with our lives."[18]

No longer in control, Fokker was suddenly forced into the role of an involved yet powerless spectator, frustrated to see revolutionary guards patroling his factory as if they owned it. As in every revolution, old scores were settled: according to Fokker, several manufacturers who showed opposition were killed. A certain amount of looting, and theft of his company's current funds, also took place, while interruption of telegraph, telephone, and mail services added to the confusion. It was impossible to ascertain what was going on in the rest of Germany.

A terrified Tony Fokker was summoned to appear before a council of workers, who demanded money and threatened to shoot him if he refused to give in to their demands. Fearing for his life, he managed to convince the tribunal that the money they wanted was in Berlin. He needed to make special arrangements, and that would take time. The council allowed him to return to his rooms for the night and appointed two men to stand guard over him and make sure he would not escape. Fokker, however, was in no mind to be led to the slaughter that easily, and with his roommate, Bernard de Waal, devised an escape plan.

That night, dressed in the uniform of my landlady's son with thousand-mark bills stuffed in my boots, I slipped out of the house past the two guards, and hurried three blocks down the street where De Waal waited in the shadows with a motorcycle. In an instant I was in the saddle and roaring out of Schwerin, headed for a village thirty miles away on the main railroad line. Through the night, afraid to show a light, I raced at forty miles an hour, fearful that at any moment I would hit something and break my neck, or run into a Revolutionary patrol.[19]

Hiding outside the railway station of Ludwigslust, he waited for a freight train headed for Berlin and managed to climb aboard a wagon full of coal. In those early days of the revolution's sweep through Germany, Berlin had remained relatively calm, and Fokker hoped he would be safer there. However, by the time the train pulled into Berlin, Germany's capital was in turmoil, like the rest of the country. On November 9, at 12:45 P.M., the news reached Berlin that Emperor Wilhelm had agreed to abdicate. All over the city red flags were hoisted, and Unter den Linden, especially the Royal Palace, became the focal point of developments. Karl Liebknecht, the foreman of the radical wing of the socialist movement and just released from prison after serving a four-year sentence for antiwar agitation, preached communist revolution after the Soviet model from the balcony of the Royal Palace. With his fellow political activist, Rosa Luxemburg, Liebknecht headed the hard core of the so-called Spartakists, a movement named after the legendary leader of a slave rebellion in ancient Rome (73–71 B.C.). They sought to cast the disjointed workers' and soldiers' councils in a Bolshevist mold. Their revolutionary seeds fell on fertile ground, for by November 1918, Berlin was a run-down city full of hungry people and with many empty shops. Digging in to defend "order" and the semblance of organized government, the more moderate Social Democrats proclaimed the republic and formed a new government headed by Friedrich Ebert, relying on the support of a changing ad hoc alliance of parties in the Reichstag.[20]

While Wilhelm II and his entourage arrived safely in Holland to be granted political asylum on the day following his abdication, Tony Fokker got off the train in the midst of Berlin's revolution. There was no question of going to his usual Berlin residence in the fashionable Hotel Bristol on Unter den Linden. Instead, he sought refuge in the house of a vaudeville girl he knew, known as Lia Lay. Though safe for the moment, his anxiety was far from over, since several half-military, half-revolutionary army units were roaming Berlin, plundering the houses of the rich. Numerous shots were fired, but this initial phase of Berlin's revolution was a surprisingly organized one, and after a while Fokker was even able to obtain a street permit, issued by the revolutionaries, through connections of Seekatz.[21] The period was extremely stressful for him, and he later admitted, "I was more affected by the horrors of the Revolution than by the whole War. They shattered my nerves. No one knew from day to day whether he was rich or poor; whether he would even escape with his life."[22]

After Fokker's hasty exit from Schwerin and his experience of hiding out at Lia Lay's, it was clear that he should leave Germany as soon as possible. With his new permit, but in the company of four specially hired bodyguards to cover all eventualities, he visited the Netherlands Consulate in Berlin on December 14 and implored the consul to give him a Dutch passport so he could move his residence back to Holland forthwith. Unprepared to deal with Dutch bureaucracy, he was told that the consul needed to check his nationality before any official document could be drawn up.[23] Expecting trouble, and determined to get out of Germany as soon as possible, Fokker decided to brave the risks, travel to Schwerin, and try to have a travel permit to Holland issued in his German passport. This proved surprisingly easy.[24] But until all arrangements for the move were completed, he was stuck in Berlin. On at least one occasion, he found himself at the wrong spot at the wrong time and had to crouch for protection in the recesses of the Prussian State Library during a shootout on Unter den Linden.[25]

In January 1919, the last real attempts of the Spartakists to seize power were brutally stopped by nationalist volunteers, united in the paramilitary Freikorps (Free Corps), who supported Ebert's government. It was no accident that Freikorps members murdered the Spartakist leaders, Karl Liebknecht and Rosa Luxemburg, on January 15, 1919. Four days later, Germany held its first postwar election, giving rise to the Weimar Republic. Nevertheless, the situation in Berlin remained unsettled, putting further pressure on Fokker to leave the country with his fiancée. Seekatz later remembered sorting out stacks of 100-mark notes on the carpet at the house of Tetta von Morgen's aunt in the chic Zoo District of Berlin and packing them to be sent to Holland.[26]

Meanwhile, in Schwerin, those remaining at the Fokker Works (now Schweriner Industrie Werke, for need of a more neutral-sounding name) were in a quandary about what to do next. Bernhard Plage, managing the factory in Fokker's absence, found it impossible to sell any of the surplus aircraft remaining on the premises.[27] Because the factory was so close to the Ostorfer Lake, the company took to building various canoe-type boats, only to find them unsalable. A second changeover, this time to commercial scales, fared equally badly, and no one quite knew what to do. In the midst of these uncertainties, Reinhold Platz tried to keep his aircraft construction team of about thirty people together. Initial efforts focused on developing a lightweight sports model of the D.VIII fighter, a project that would soon founder: no customers in Germany were willing to spend money on such an expensive hobby.[28]

Besides, Tony Fokker, resolved to cut his losses and move to Holland, had altogether different plans. Communicating by telephone with Heinrich Mahn, his chief of shipping in Schwerin, Fokker devised a grand operation using the confused situation in Germany as a cover to save Schwerin factory equipment and stocks, and move the lot by train to Holland. In retrospect, even Fokker himself found it difficult to believe that the plan had actually worked.[29] It relied on the fact that early in 1919 the situation in Germany was still very unstable. Regulations on the export of goods from those zones of Germany not yet controlled by the Allies to neutral countries like Holland did not exist. Nearly all of the measures having to do with import and export concentrated on the transport of goods *into* Germany from or through the neutral countries. Exports were, at that point, only loosely monitored, though Allied policy stated that no German goods could be transported through neutral countries without special permits. These measures aimed to pressure Germany into full acceptance of the Allied peace terms. During the first months of 1919, most of the Allied naval blockade of the North Sea remained intact, and the Supreme Economic Council monitored the movement of goods into the ex-enemy and neutral states through its Inter-Allied Trade Committees.[30] This monitoring was somewhat haphazard because the Inter-Allied Trade Committees did not always know of the decisions of the Supreme Economic Council and therefore could not carry out their task properly.[31] On the other hand, the Dutch followed their own policy regarding imports from Germany, and the government consciously turned a blind eye to such activities. To the British ambassador in The Hague, Robert Graham, the Dutch foreign minister, Herman van Karnebeek, stated that his government could not be expected to supervise all imports from Germany by private enterprises and individuals, especially because it was often "difficult to ascertain which of those imports are war materials."[32] Capitalizing on this attitude, a number of key firms from the German armaments industry transferred parts of their factory equipment, and their assets, from Germany to Holland and other surrounding neutral countries in semiofficially condoned efforts to sabotage the dismantling of Germany's war industry by the Allies.[33] Later that year, Fokker's former partner Junkers, for example, had twenty-four aircraft crated and stored across the Dutch border in the city of Nijmegen.[34]

Fokker's plan, which he called "the great smuggling plot," thus fit into a larger pattern of exports of military equipment from Germany and was neither as unusual or illegal as he would have liked people to believe.[35] The operation was very carefully planned and probably prepared at three different locations: Berlin, Schwerin, and Holland. In Berlin, Fokker and Wilhelm Horter, who managed Fokker's Berlin-based Zentralbüro, approached various authorities in the Reichsverwertungsamt (State Valuation Bureau), a government branch specially established to sell off surplus war material, to ask their cooperation. As an official representative of the Reichsverwertungsamt, none other than Hermann Göring, then a lieutenant, was sent to Schwerin with another official and Seekatz to arrange the financial terms under which Fokker would be allowed to buy back aircraft and engines sitting on his premises—material already paid for by the military, but undelivered because of the end of the war, for which Göring offered Seekatz an extremely good deal.[36] As a result, Fokker bought back 92 D.VII fighters and 100 rotary engines.[37] Trusted factory employees removed aircraft and factory equipment from the view of representatives of the Inter-Allied Aeronautical Control Commission, which was supposed to supervise the dismantling of Germany's military industry. In Schwerin Heinrich Mahn set about "organizing" railroad cars, which were short in supply, and preparing the actual transport. Finally, after arriving in Holland on his German passport with Tetta on February 16, 1919, Fokker managed to hammer out a deal with the small Trompenburg car and aircraft factory in Amsterdam. Trompenburg agreed to act as the official importer of the Fokker shipments.[38]

All had been arranged when the first of six trains, 350 rail cars in all, crossed the Dutch border at Oldenzaal on March 18, 1919. According to Fokker, they carried 220 aircraft (120 D.VII's, over 60 C.I two-seater observation planes, and a number of D.VIII's and other planes), some 400 aero engines, and an important part of his factory stocks and equipment from Schwerin to Amsterdam. Ingeniously, Mahn had arranged for each train to consist of about sixty cars, making it necessary for the trains to stay on the mainline track at the Hannover railroad junction and minimizing the risk of their being stopped and inspected on the way: the sidings at the frontier station of Bentheim were also limited to a capacity for trains of up to forty cars. Because it was not possible to run a locomotive out of Germany, an arrangement was made to have Dutch locomotives brought up the line to Salzbergen, the last station in Germany that the trains passed—arrangements with the Dutch railways that may have been facilitated by the fact that Fritz Cremer's father, Jacob Cremer, had, until a few weeks before, held the position of chairman of the Joint Boards of Control of the two railway companies in Holland.[39] Border checks by Dutch customs were minimal, and the trains safely rolled into neutral territory to be parked at the shunting yards near the Amsterdam Petroleum Harbor until Fokker had decided what to do with them.[40] By this careful organizing, the trains escaped being stopped by the Allied Control Commissions. Indeed, the shipments were carried out with such a low profile that the files of the United States Legation in The Hague on trading with the enemy and war contraband did not even mention Fokker.[41] Not until a year later did information relayed by British Military Intelligence to the Americans read, "It has been reported that Fokker, the aviator, has brought out a large number of aeroplanes from Germany and painted over German marks, in contravention of the Armistice terms."[42]

If Fokker's exit from Germany went virtually unrecorded by the Allies, his departure did create something of a stir in Germany when it became known. In the popular press, a cartoon depicted Fokker merrily flying away in his private plane, towing a bag of money with "100 million marks" written on it. Though this was an exaggeration, he did carry off a sizable part of his fortune in a curious assortment of currencies and banknotes. In *Flying Dutchman,* Fokker describes how he motored to Travemünde on the Baltic shore with a carful of suitcases and bags of money totaling several million marks and a few hundred thousand dollars worth of coupons. All was brought aboard his 28-meter powered sailing yacht, *Hana,* which he kept moored there, and its captain sailed the ship to Holland, delivering the precious cargo safely to an anxious Fokker. Another $400,000 in foreign currency was packed into an old suitcase belonging to a cook from one of the foreign diplomatic legations in Berlin who traveled to Holland on the same train as Tony and Tetta did. In all, Fokker estimated that he saved 25 percent of thirty million marks in profits he accumulated in Germany during the war. The rest, according to Fokker again, was partly lost in unlucky investments and in banking deals that did not materialize, or was seized by the German government, which assessed his accumulated personal tax liabilities at 14,251,000 marks. He refused to pay, and the Germans threatened to confiscate his properties in Germany. A negotiated settlement was later reached, reducing his total accountability to around six million marks.[43]

## GERMAN OR DUTCH?

On February 16, 1919, Tony and Tetta arrived in Holland. For both of them this meant a serious adjustment, and not only because they initially moved in with Tony's parents in Haarlem. On the threshold of a fresh start, the couple had imagined they would be married in the peaceful surroundings of Haarlem as soon as possible and take life easy for a bit. Tetta had obtained a marriage license in Berlin on January 14, though she knew her father could not be present because his military status still prevented him from crossing the border. But when Fokker visited the town hall, assuming he would be able to obtain a marriage license easily, he discovered himself the subject of a legalistic controversy concerning his nationality.

In accordance with the 1892 Law on Netherlands Citizenship, the Dutch consul in Berlin had asked the mayor of Haarlem, Fokker's last place of residence in Holland, to issue an official "statement of citizenship," which he needed to give Fokker a new passport. Such a statement required a prior investigation into the antecedents of the person involved.[44] The mayor, however, found it impossible to cooperate because he had heard rumors that Fokker had been naturalized as a German citizen. Fokker had not reckoned on such problems arising and found this development highly puzzling. Though he had lived in Germany for eight years, he had always considered himself Dutch. In December 1914 the naturalization issue had been a bit of a nuisance, more a matter of maintaining his position as an arms producer in wartime Germany than one of deliberate choice: it had simply seemed opportune to cooperate with the authorities. Now the question of his nationality took on an entirely new and unsettling dimension. With Germany defeated and in the midst of a revolution Fokker wanted to escape from, it would be far preferable to have a Dutch passport and be able to cross the border without special visa requirements. Fokker denied having taken German citizenship of his own free will, insisting that he had been forced to cooperate in the matter and had never formally recognized his new nationality. Early in January 1919, the consul approached the Kriegsministerium, the German War Office, at Fokker's request to seek some sort of solution. With various government agencies already involved in moving vital parts of the German military-industrial complex out of the country, and keeping a benevolent eye on Fokker's preparations in this respect, the Kriegsministerium was quick to declare that Fokker's naturalization had, indeed, been achieved by coercion and that Germany would give up any claims on Fokker's citizenship if Fokker so desired.[45]

The mayor of Haarlem, however, would not budge, insisting the law clearly held that Dutch citizenship had been lost through naturalization. It said nothing about the use of force, and the mayor had no intention of overstepping the boundaries of his authority, though he was prepared to listen to Fokker's argument after his arrival in Holland. In the end, the mayor put the decision in the hands of his superior, the queen's commissioner for the province of North Holland.[46] A frustrated Tony Fokker then sought top-rate legal council by engaging Professor A. S. Oppenheim of the prestigious Law Faculty of Leyden University, an acquaintance of his Uncle Eduard, to find an opening in the legalistic facade of Dutch officialdom. While Oppenheim tried to soothe Fokker on the basis of previous jurisprudence, the latter took recourse to his family's social connections to get what he wanted in a more roundabout fashion, going as high up as Queen Wilhelmina's husband, Prince-Consort Hendrik, Duke of Mecklenburg-Schwerin and patron of the Netherlands Royal Aeronautics Society. The question made it all the way to Prime Minister Charles Ruys de Beerenbrouck, who also held the portfolio of Internal Affairs.

After receiving assurances that the nationality question would soon be settled, Tony and Tetta, still living with Tony's parents in Haarlem, decided not to postpone their wedding any further. On March 25, 1919, they were married in Haarlem's town hall in a secluded and unobtrusive civil ceremony. Fokker's friend Frits Cremer acted as a witness, together with another friend, Geert Hoekstra, a physician from Utrecht.[47] However much Tony Fokker liked the attention of the press, he was always very particular about keeping reporters away from the details of his private life. In one of the rare surviving photographs, the newlywed couple are seen descending the stairs of the town hall after the ceremony in chic but plain clothes. Tetta did not even wear a bridal dress and had to be satisfied with a few white carnations on her coat. Tony's father could hardly wait to light his cigar until the small gathering of invited guests had returned to the parental house in the Kleine Houtweg. Prince Hendrik, an acquaintance of Tony and of Tetta's father from prewar Schwerin, was among those who came to wish them well.[48]

That very same day, by voice of Minister of Justice Theodoor Heems-kerk, the cabinet finally decided to waive the legal objections and instruct that a passport be issued to Tony Fokker. This decision, it was whispered, owed much to a personal intercession by Prince Hendrik.[49] The favor was a personal one and did not extend to Tetta Fokker–Von Morgen. It also reflected the government's high expectation that Fokker would be instrumental in bolstering Dutch neutrality by providing Holland with its own indigenous aircraft industry.[50]

If this initial welcome in Holland was rather cool, Tony Fokker was soon to be embraced by the nation as a lost son come home. In a poll organized by the popular weekly *Het Leven* in December 1922 to name the twenty most popular men in Holland, Tony Fokker ranked fourth after musician Willem Mengelberg, Foreign Minister Hendrik van Karnebeek, and actor-entertainer Louis Bouwmeester.[51] He was, and is to this day, publicly acclaimed as one of the Dutch national heroes, the subject of a full-length movie in 1957, *De Vliegende Hollander*,[52] and honored by his inclusion in various national biographic dictionaries.[53]

After his marriage, Fokker allowed himself a few months to rearrange his life. To someone who had never lived in the country except as a schoolboy, the sedate and scrupulously administered life in Holland seemed as oppressive to Tony as it was foreign to Tetta. Feeling out of place, they doted on the romantic notion of taking Tony's three-master schooner, which was waiting for them in Denmark, on a three-year extended honeymoon trip around the world. Nothing came of it. Instead, Fokker soon found himself engulfed in business again, planning to start anew in Holland.[54] Perhaps to compensate for the trip, he later named the luxurious 82-foot motor yacht that he bought *Honey Moon*.

For the first month of their marriage, the newlyweds boarded in the expensive Hotel de l'Europe in Amsterdam. Late in April 1919, they moved to a stately house in that city's Roemer Visscherstraat: a detached, spacious, gray-plastered, nineteenth-century building with large windows. There, overlooking a semienclosed garden, Tetta spent many an hour engaged in her favorite hobby, needlework. Her pride was a large Gobelin displaying the heraldic weapon of the Fokker family, which she worked on for several years.[55]

Tetta saw much less of her husband than she had expected, and their dream of sailing the world soon evaporated. In the first week of May 1919 Fokker opened a new office on Amsterdam's prestigious central street, the Rokin. Reminiscent of his days in prewar German, he poured his energy into establishing an aircraft factory with a flying school attached to it. He was determined to start anew in aircraft manufacture and with a clean slate, detaching himself from his German past. For this reason he was careful to place his shares in the Fokker/Schweriner Industrie Werke GmbH in the care of the Unie Bank voor Nederland en Koloniën, thus creating a legal buffer between himself and his German companies, lest there be liabilities chasing him across the border.

## GERMAN SHUTDOWN

The departure of its owner-director did nothing to improve the fortunes of the Schweriner Industrie Werke (SIW). Plage left the company, and in the absence of Fokker, the management of the works came to rest jointly on the shoulders of Hans von Morgen, now Fokker's brother-in-law, and chief engineer Curt Stranz, both of whom held powers of attorney. Reinhold Platz continued to run aircraft production, such as it was. Friedrich Seekatz went away to join Germany's small airline, Deutsche Luft Reederei (but returned to Fokker in 1921). In Berlin, Wilhelm Horter and Fokker's personal secretary, Maria Moslehner, stayed behind to run the remnants of the Fokker-Zentralbüro, now moved from the fashionable Hotel Bristol to a small office in the Charlottenburg suburb. Officially an export company for all sorts of goods, the Fokker-Zentralbüro soon took as its main task the scaling down of Fokker's German interests, while acting as a buffer for various legal and tax liabilities.[56] Fokker's other Berlin-based venture of the months immediately after the war, Luftbild GmbH, which sought to promote aerial photography, was also short-lived.

In Schwerin, business was slow, despite continued efforts to build small motorboats and production of subcomponents for cars and railway stock, and the company soon slumped into decline. The work force was cut down from 680 in April 1919 to 140 in June 1921.[57] Attempts to re-export the aircraft that Fokker had taken with him to Holland and perhaps even reinitialize production of a limited series of D.VII planes for export to the Scandinavian countries (through the activities of special sales agent and flying ace Hermann Göring) failed because Fokker was obviously not interested in keeping the Schwerin factory open. He had long since reached the conclusion that his fame, and the special mention his products had received in the Armistice Agreement, doomed the continued manufacture of Fokker aircraft in Germany. From Copenhagen, Göring complained several times to an unresponsive Fokker that he had staked both his name and his own money obtaining some orders, and urged Fokker to come forward with extra cash that would enable him to persevere and close a deal:

> All I ask for is a regular daily allowance, and a percentage for each plane that I sell here. In the meantime I do risk my neck and my bones in these demonstration flights. Be assured that I am very busy on your behalf. . . . For that reason I beg you again not to let this matter drag on, but to tell me your decision in this respect. You will understand that I would like to know where I stand. Excuse me for being so open and explicit about everything.[58]

In spite of Fokker's lack of interest, aircraft construction was kept alive in Schwerin, albeit at a slow pace, with Platz, the oldest employee, in charge. In December 1918, Stranz's engineering department had come up with an audacious new idea: what if they were to take the basic design of the latest D.VIII fighter and enlarge it to carry passengers? The idea probably emerged from an aborted attempt to develop a plane to Idflieg specifications for a long-range reconnaissance aircraft *(Fernaufklärer)* in October 1918. It formed the basis of the V.44 (F.I) project of 1919: a single-engined airliner with six seats paired side by side in an open cockpit. When the airplane was nearing completion, the project was abandoned in favor of an improved version with an enclosed cabin for four passengers, the V.45 (F.II). The prototype made its first flight from the Görries airfield near Schwerin in October 1919. Tony Fokker had not been there to see the aircraft. The only time he would return to Schwerin was for one short and final visit on his birthday, April 6, 1923, at the height of Germany's inflation crisis, with Seekatz. His formerly proud establishment looked dilapidated and had nearly ceased to exist. After Hans von Morgen started liquidation proceedings on May 12, 1922, barely a handful of people were employed there any more.[59] The SIW finally closed down in 1926. By then Fokker had disposed of most of his former German assets.

# 4

## A BUSINESS OF SORTS

Peacefully neutral in the four years that much of Europe burned, Dutch society, and its ways, had maintained many of its nineteenth-century characteristics. Historians have cast Holland both as "a conservative country"[1] and as "a liberal country."[2] Whereas social and political unrest swept through Germany in November 1918, such developments in Holland remained limited to a single gathering of the Social Democratic Party in Rotterdam, an outcry in parliament, and a single casualty during a march in Amsterdam. Instead, tens of thousands of people gathered on the Malieveld (the Common Green) in The Hague, a spontaneous demonstration of loyalty to the queen on November 18. It was logical, then, that social and political developments followed sedate patterns, regulated by coalition governments that tended to have a high representation of technocrats and practiced policies of accommodation rather than of confrontation. Dutch society continued to function according to a peculiar mixture of religious divisions and strictly hierarchic patterns, with much power vested in the bureaucracy. In the economic field, policies of laissez-faire prevailed, though Dutch capitalism was gentle in character. Big business expressed its influence in smoky salons rather than in overt shows of financial power.

At the time of Tony Fokker's arrival, Holland possessed a tiny aircraft industry of its own, built up with the support of the government during the war years to ensure the continued supply of aircraft for the country's defense needs. Yet neither the army's Air Corps (Luchtvaart-Afdeeling, founded on July 1, 1913) nor the Naval Air Arm (Marine Luchtvaart Dienst, founded on August 18, 1917) was of much consequence. A reluctance to spend significant sums of money on these new branches of the armed forces had kept them very small: in 1918 the Air Corps' official strength stood at some 135 aircraft, though the majority of them were barely operational, having been acquired through internment; and Naval Air Arm strength at about 37 aircraft.[3]

In 1916 a small Dutch automobile factory in Amsterdam, the N.V. Automobiel Maatschappij Trompenburg, which made cars under the brand name of Spijker (nail), had diversified into the aeronautical field. Because the company's managing director, Henri Wijnmalen, was one of Holland's earliest pilots, this was not entirely surprising. A separate division of Trompenburg was created to begin license-manufacturing airplanes for the Air Corps. Apart from Trompenburg, a Rotterdam-based factory of retail weighing and cutting instruments, N.V. Maatschappij Van Berkel's Patent, had secured a contract from the Naval Air Arm to supply forty-two floatplanes of its own design in the week before the signing of the Armistice Agreement. For this project it had engaged the services of Albert von Baumhauer, a graduate in mechanical engineering of the nearby Technical University of Delft.

With his customarily keen sense for commerce, Tony Fokker realized he would have to come to some sort of cooperative understanding with the existing producers if he wanted to set up his business in such a small country as Holland. Taking up temporary residence at his parents' house, he lost no time getting together with Wijnmalen to hammer out a deal involving Trompenburg. Fokker offered to take over Trompenburg's Air Corps contract for 98 fighter aircraft and 118 reconnaissance planes. He needed the business badly, because the airplanes and factory inventory that he had managed to railroad out of Germany were sitting idle on their wagons on the rail shunting yard of the Amsterdam Petroleum Harbor. Knowing this, Wijnmalen was in no hurry to reach an early accord, seeking to extract the highest possible price from Fokker. It was essential to drive a hard bargain, since a Fokker factory would be a lethal threat to Trompenburg's own aviation activities.[4] Not until January 2, 1920, did Wijnmalen, Fokker, and the Ministry of War reach an agreement under which Fokker would deliver 92 D.VII and 92 C.I aircraft to the Air Corps. Yet four months later, the government Munitions Bureau, facing new cutbacks in military spending, changed its mind. Trompenburg was bought out of the contract, and Fokker had to settle for an order of 60 C.I reconnaissance aircraft and 20 D.VII fighters for 600,000 guilders.[5] Three years later, Trompenburg went into liquidation.

To make renegotiating the Trompenburg order more acceptable to the Dutch government in the context of postwar Holland, Fokker realized that it would help him to found a Dutch company. After all, the very reason the government had for doing business with Trompenburg was that a neutral "middle power" like Holland needed to be independent of foreign suppliers of arms and military equipment. Forgetting about his intended life of leisure and honeymooning, Fokker made his first move. On July 2, 1919, using aircraft brought in from Germany, he organized a small barnstorming operation on the Scheveningen beach, a mere three miles from where the government met with parliament in The Hague, under the name of Fokker's Luchttoerisme (Fokker's Air Tourism).[6]

Fokker understood that his German past would not be easily forgotten and that a serious new venture should, above all, strongly emphasize being Dutch. Again his family stepped in to help. Through Jacob Cremer he was able to enlist the support of the prestigious national celebrity, Gen. Joan van Heutsz, former governor-general of the Netherlands East Indies. Van Heutsz, a radical nationalist, was brought in to act as chairman of the Board of Control of a new limited-liability company. Other members of the board reflected the circle of people Fokker had been in contact with since his arrival in Holland: Cornelis Vattier Kraane, director of the Amsterdam-based trading company Blaauwhoedenveem Vriesseveem; Frits Fentener van Vlissingen, a well-known millionaire-entrepreneur and corporate investor, who also took part in the founding of KLM that year; Johan van der Houven van Oordt, the former head of the Ministry of Colonial Affairs; Daniel Patijn, co-owner of a stockbroking firm; and Tony's uncle, Eduard Fokker. Careful to avoid the name Fokker for obvious international political reasons, the Nederlandsche Vliegtuigenfabriek NV was officially founded before a notary in Amsterdam on July 21, 1919. The Unie Bank voor Nederland en Koloniën and a stock-brokerage firm, Patijn, Van Notten & Co., fronted as the two founders, each participating equally in the company's shareholding with investments of 275,000 guilders. Tony Fokker, discretely absent at the signing of the official papers, was only mentioned as the company's first managing director.[7] The maximum number of shares to be issued by the company was put at 1,500,000 guilders, and the purpose of the company was stated as the production of, and trade in, aircraft, and fast-running boats, and parts thereof, and everything that is connected with such;

the foundation and operation of flying schools, airports, landing strips, and repair workshops, either under its own account, or on the account or with participation of third parties, and for organizing and providing airshows.

The operation of traffic in the air for persons, letters, goods, and so on, the provision of aerial photography for exploration and for topographical purposes, either under its own account, or on the account or with participation of third parties.[8]

For all of the Board of Control's expertise and renown, its influence on the company was negligible. In the months following the founding, Fokker reacquired practically all the stock of the Nederlandsche Vliegtuigenfabriek, soon to be known in local parlance as the Fokker Factory, giving him full control over his business. This would have serious repercussions for the firm's further development. Fokker's approach to commerce was often haphazard: he traveled around a lot and was frequently absent from his office. Over time, his corporate and private capital became increasingly intermingled, and his financial manager, F. Elekind, had a full-time job keeping track.[9] Though Fokker put himself on the payroll for a modest director's salary of 18,000 guilders annually, he claimed over 50,000 guilders in expenses each year. Most of these were travel-related, for he spent a lot of his time away from the factory on business trips.[10]

Sixteen days after the founding of the Nederlandsche Vliegtuigenfabriek, Fokker was present at the biggest aviation spectacle in Holland for decades to come: the First Amsterdam Air Transport Exhibition, known by its Dutch acronym, ELTA (Eerste Luchtverkeer Tentoonstelling Amsterdam). The show was officially opened by Queen Wilhelmina herself. The ELTA had been initiated, and then organized, by two lieutenants of the Air Corps, Albert Plesman and Marius Hofstee, who had managed to enlist significant corporate financial support, funds from the city of Amsterdam, and even from the government. ELTA festivities went on for more than six weeks, drawing more than half a million paying visitors from around the country to the soggy pastures north of Amsterdam where the spectacle took place. For Fokker, it was an ideal opportunity to demonstrate his flying skills and machines before a national audience and at the same time leave his calling card with the Dutch military authorities. Tetta tagged along on several of his demonstration flights, which left her pale and shaky, looking slightly out of place in her thick leather jacket and flying cap in the few surviving photographs, yet casting a timid smile toward the camera as her husband was laureled for his performance.[11] He took his parents up with him, too.

Fokker's connection to the exhibition would be a lasting one: after it closed he bought its buildings to convert them into a new factory. The grounds were rented from the city of Amsterdam, and company activities there began on September 16.[12] The location was hardly suited for aircraft production because the factory grounds had no landing strip. Throughout Fokker's life, all aircraft had to be shipped to Schiphol Airport by river barge for final assembly in Fokker's hangar before they could be test-flown. Yet the ELTA buildings came cheap, and that was their main attraction. Fokker also rented temporary extra floor space to produce floatplanes for the Naval Air Arm at the unused naval yard of Veere, in the southwest part of the country, from 1921 to 1927. This rather odd organization was typical of Fokker's financial management. Perhaps weary from the initial years of financial hardships, and overly cautious after the loss of his German company and possessions after 1918, Fokker conducted a very parsimonious financial policy, avoiding long-term investments.

After the first batch of aircraft was delivered to the Dutch Air Corps, Fokker was still stuck with about 140 of the airplanes he had brought across the border from Germany. As a stop-gap measure, they were stored in a warehouse in the Amsterdam Harbor, while he set about organizing his new company. This took a considerable time, especially because Fokker himself had no clear-cut idea of what he wanted. In January 1920 the work force totaled eighty-eight.[13] A number of people from the upper echelons of Fokker's German factories and the Fokker Zentralbüro in Berlin were transferred to Holland and taken on board as staff by the new company. The most influential of those early arrivals was Fokker's Zentralbüro director, Wilhelm Horter, who became general plant manager in Amsterdam. From Schwerin, Fokker's friend Bernard de Waal arrived to become his new chief test pilot and supervisor at the flight-testing facility at Schiphol Airport. He remained with Fokker until his untimely death at the age of thirty-two, after a short illness, in July 1924. Reinhold Platz was brought to Holland as Fokker's chief constructor in 1921. This was a fateful decision, for Platz had reached the limits of his abilities by the early 1920s. Described by Fokker's deputy director from 1925 to 1935, Bruno Stephan, as a narrow-minded man, who, like

Fokker, had very little knowledge of or regard for aerodynamic theory, Platz flatly rejected new ideas that went beyond his understanding. Stephan later remarked, "He never even took the trouble to learn the language of the country he worked in, hardly ever read, nor studied that what even he could still have grasped, despite his faltering education, and his knowledge, therefore, depended on what he had taught himself, which showed to be insufficient in the long run."[14]

To extract at least some use from his unsold stock of D.VII and D.VIII fighter aircraft, Fokker had Horter pursue various foreign contacts to dispose of them. Exports, Fokker realized early on, were going to be vital to the success of his company in a small country such as Holland, which could offer only a limited home market. When foreign sales initiatives did not produce any results, Fokker came up with the idea of converting the aircraft to fast sports and aerobatics planes. Nothing worked until, finally, Friedrich Seekatz was brought in as an aircraft salesman in June 1921.

Meanwhile, in December 1919, rumors reached Fokker that the newly founded KLM was considering starting a flying school at the Maaldrift airstrip near The Hague. He immediately offered to step in on a joint-venture basis. The bid came to nothing because KLM's management decided that its real future would lie in air transport, rather than air training.[15] A month later, KLM, which did not yet have any aircraft of its own, also declined to charter Fokker's airplanes for air taxi operations, fearing political repercussions if they landed these German machines abroad.[16] For similar reasons, the Dutch postal service turned down a proposal for a Fokker-operated airmail service to Britain.[17] The needs of the Air Corps, so far Fokker's best Dutch customer, had also been completely satisfied. Though the Air Corps expressed some commitment in encouraging Fokker to stay in Holland, it had neither the money nor the need for more airplanes.[18] By July 1920, it had become quite obvious that the growth of his new company depended entirely on exports, an area in which Germany would continue to occupy a position of importance.

## GERMAN ECHOES

After four years in the forefront of German aeronautic development, it took Fokker a while to shed his ties with the past. Several burdens of his German life were still difficult to shake off. To start with, the legal controversy over the Schneider patent on the synchronization of the aircraft machine gun was far from settled. Though Fokker had gotten nowhere in his efforts to have the negative verdict of the Königliches Kammergericht (June 30, 1917) in Berlin overruled, he had time on his side. Late in 1918, before moving to Holland and seeking to rid himself of the lawsuit, Fokker had passed on the rights of his own synchronization mechanism to the Schweriner Industrie Werke. The factory was headed by Tony's brother-in-law, Hans Georg von Morgen—a typical Fokker appointee, within the circle of family and friends. His job was an unenviable one: gradually scaling down activities and leading the company into liquidation, because it seemed clear that Germany would be wiped off the map as an aeronautical power.

At first the SIW continued to build airplanes. Fokker's first two commercial models, the F.II and F.III transports, were built there in 1920 and 1921 under Platz's supervision. Then the Allied Control Commission zeroed in on the former Fokker plant, and it ended its active days building small, high-powered boats. After Von Morgen started liquidation proceedings in May 1922, however, Fokker was in no hurry to finish them. He could still use the hollow shell of the SIW as a shield against indemnities that might be filed against him as a result of the Schneider lawsuit.[19]

The lawsuit itself continued to develop badly for Fokker and the SIW. Round upon round of the legal battle had been lost, and as a result, claims for damages were staggering. Yet Fokker, honestly believing the invention to be his own brainchild, and one that nobody could take away from him, told his team of lawyers to pull out all the stops and pursue the case no matter what. Drawing on his lavish finances, he was able to do so, while his opponents, the LVG company, itself having gone into liquidation in 1922, had to resort to such odd measures as seeking a court order to seize three oak tree trunks from SIW to pay the bills during Germany's hyperinflation of 1923, when the mark collapsed. Fokker's defense that he and SIW were no longer identical held up in court, and the seizure came to naught.[20] Nevertheless, pressing on in the case, LVG had Fokker sentenced on November 16, 1923, to pay 1,000 goldmarks for infringement of patent rights before he had transferred the rights for his own invention to his company.[21] As before, Fokker flatly refused to pay (or rather, did not answer at all), thus forcing LVG to start yet another legal proceeding to extract the money.

This development led to the breakdown of the basic case when

Fokker's lawyers persuaded the court to reopen investigation of the validity of the LVG's patent claims as such. In 1926 the receivers of the now-dissolved LVG proposed a settlement of 150,000 marks. Fokker's remaining lawyer on the case, Hugo Alexander-Katz, seconded by Von Morgen, advised him to go along with it.[22] Fokker would not hear any of it; in his view, paying would have been an admission of guilt, where he felt there simply was none. On December 14, 1926, the Berlin court gave off a provisional verdict: Fokker owed LVG 50,000 marks for infringement of the Schneider patent between September 1915 and June 1916.[23] Once more, Fokker refused to pay. Meanwhile, the receivers of LVG had decided to give up the case the month before, because there was no way Fokker could be made to pay within a reasonable time.[24]

Having held his breath longer than anyone, Tony Fokker must have been relieved to hear the news after ten years of legal battle. Yet the Schneider patent came to haunt him once again, in 1933. The unfortunate Franz Schneider, then aged sixty, had by that time become a victim of the general economic crisis. It had been several years since he last held a steady job. As a consequence, he had slid down the German social ladder, and if that was not bad enough, his health also started to fail. In April 1933 several of his old friends told him that the recently published German translation of *Flying Dutchman* told Fokker's version of how he came up with the synchronization mechanism without even mentioning Schneider. Outraged at not having his part in the invention recognized, Schneider, who had salvaged his old patent from LVG's liquidation, tried to sue again—not so much for financial damages, this time, but in a belated attempt to receive recognition in Fokker's book. Schneider was resentful of his fate in life, and, like so many others, firmly entrenched in the Nazi movement as a result. He found a lawyer, Erwin Graf, who was willing to represent him free of charge under the provisions of the German poor law, "even though this lawyer was a non-Aryan, yet admitted by the present Government, because he had fought in the war as a soldier and earned the Iron Cross, and the Cross for the Wounded. He had repeatedly been badly wounded during the war. He had a friend, however, an Aryan lawyer, who regularly represented him."[25]

A very annoyed Anthony Fokker immediately put his lawyers on the case to stop Schneider dead in his tracks, yet admitted in a letter to his Amsterdam law firm that it had been legally ruled that his invention was covered by the Schneider patent—the only time he made the admission.[26] While denying that Fokker's book, to which he did not own the copyrights (he had granted them to the actual writer, Bruce Gould), could be detrimental to Schneider's interests after so many years, Fokker's lawyers successfully focused on the poor-law protection Schneider sought. On September 28, 1933, the Berlin Landesgericht (District Court) dismissed Schneider's case because it had no real prospect of success. Poor Schneider was even condemned for trying and saddled with the legal charges of the court: 1,000 marks. Fokker had won. After the most intense of his legal battles, the invention was his.

The Schneider case was not the only legal battle that pursued Fokker from Germany. Between 1923 and 1939, he was on the defensive in a series of highly technical patent cases brought by Junkers over the construction of internally braced cantilever wings.[27] Tax issues also chased him across the border. In *Flying Dutchman,* he described how he had to settle his German income and property taxes during the months after his arrival in Holland. In 1919 the Schwerin tax commissioner appraised his overdue tax payments at 14,251,000 marks. To secure payment the German authorities confiscated all of Fokker's possessions.[28] After some negotiating a settlement was reached, whereby his liability was reduced to six million marks, to be paid over a five-year period.[29]

Besides that, Fokker stood accused by the Reichsmilitärfiskus (the military branch of the German internal revenue service) of embezzling four million marks between 1916 and the end of 1918.[30] Like most countries at war, the German government had issued war bonds to collect extra revenue. To encourage the sale of these bonds, the government allowed both individuals and companies to apply for a cash advance to pay for them. The advance itself came in the form of a loan that remained free of interest for up to one month, after which it had to be converted in full into war bonds. Like all loans, these advances had a price tag attached to them: interest varied between 4 and 5 percent. Because the bonds themselves also yielded 5 percent, investment in them was praised as a patriotic act, not as a secure investment. In September 1916 Fokker asked for such an advance of one million marks from the military authorities. A second advance of three million marks was made available in June 1918. But Fokker did not convert the four million into war bonds, as the military tax service concluded when examining the case in December 1920. Where had the money gone, and why had Fokker never paid any of the 252,500 marks in interest the loan should have brought in?[31]

The case surfaced only after Fokker had left Germany. On May 31, 1919, he sent an employee, Otto Borde, the assistant sales manager of

Fokker's Zentralbüro in Berlin, with 4,047,000 marks worth of war bonds to the military tax inspector to repay the four million he had been advanced in 1916 and 1918. But because there was no record that Fokker had converted his advances into war bonds, Borde's visit raised suspicions: where did these certificates come from? Besides, as the tax inspectors pointed out to Borde, the value of the certificates had declined. After all, the war had been lost, which made the bonds an even worse investment, and the true value of the package was only 3,966,060 marks. Borde thereupon pointed out that not all the interest coupons that came with the bonds had been cashed, and that the accumulated value of those coupons was 33,725 marks, narrowing the balance due to a mere 215 marks. At a loss how to handle the situation, the tax inspectors knew no better than to accept the bonds. Their superiors, on the rebound, took a rather different view: with no record of Fokker's converting his advances into war bonds, they took the position that, apart from the fact that their total value now fell short of his debt, Fokker could not now present bonds to repay the loan, and that since the advance had been paid in cash, Fokker had to repay in cash also. Furthermore, the state had never received any of the 252,500 marks in interest. On the contrary, instead of paying interest, Fokker appeared to have collected a dividend on the bonds! The military authorities were outraged at such "unpatriotic behavior."[32]

Sorting out what had actually happened took a long time, and not before June 1922 did Fokker's tax lawyers complete their defense. They held that the advances had never been intended to serve as a means to invest in war bonds: By 1916 the administrative framework of the armed services had become so sluggish that it failed to pay Fokker on time for the delivery of aircraft. So the only way he could meet immediate cash-flow problems was to show an interest in buying war bonds to obtain a quick cash advance. It was understood at the time that the advances were related to delayed payments, not to the war bonds scheme, and would be free of interest, even though the official papers referred erroneously to them as loans toward the purchase of bonds. To substantiate their claim, Fokker's lawyers even produced a witness, an ex-clerk of the military payments office, who was prepared to speak out in court. And if accepting the war bonds as a form of repayment for the cash advance had been a mistake (which Fokker's lawyers of course denied), the costs of such a mistake should be borne by the state.

Around these central arguments they laid a thick screen of legal smoke

that obscured one of the central points in the case: was the cash advance
that Fokker received really interest free, and had Fokker been within his
rights to hang on to the money for such a long time? It did not say so in
any of the documents.[33] Digging to strengthen their case further, Fokker's
lawyers unearthed an unpaid bill for 350,000 marks worth of factory-
owned war materials destroyed in 1919 at the Fokker factory in Schwerin
on orders from the Weimar government.[34] With the issues now becoming
more and more complex, and the German mark plummeting in value
in 1923, the military taxation authorities proposed to drop the case,
provided the SIW abandon its claim for the 350,000 marks. The state offered to pay 91 percent of SIW's legal fees. Final settlement was not
reached until September 1926, because the payments had to be converted
into the reconstructed German currency unit, the reichsmark.[35] Thus, by
the end of 1926, with the SIW liquidated, the Schneider case seemingly
buried, and financial matters settled, the German chapter in Fokker's life
finally closed.


## SALES FAR AND WIDE

By 1921 Tony Fokker had still not been able to dispose of the surplus
D.VII's from Germany. The matter bothered him, because renting storage
space cost money, even though the aircraft themselves should have been
written off long before. He sent his new export manager, Friedrich
Seekatz, on a mission to find a buyer for the remaining D.VII's. At his
best in such shady deals, Seekatz reported back that he had finally found
a purchaser for the airplanes: the Soviet Union. The pariah of Europe,
banned from the international arena since 1917, Lenin's communist
regime was about to become Fokker's best customer. Absolutely uninterested in politics, and in the possible embarrassment that his exports to the
Soviet Union might present to the government of the Netherlands, Tony
Fokker left the Soviet contacts to Seekatz and his assistants.

Following Seekatz's initial arrangement with a Russian middleman
named Sjirinkin, the Fokker company exported fifty of its remaining
stock of D.VII fighters and three C.I reconnaissance aircraft to Moscow
in 1922. Records (when kept) were always sketchy, and different sources
later quoted different figures, but according to a senior official at the
Netherlands Ministry of Foreign Affairs, exports to the Soviet Union
amounted to 338 military aircraft in 1923 alone, at a cost of 1,811,000

guilders.[36] Business was booming, and though officially rejecting the atheist Soviet regime, the Netherlands government played an active role in facilitating Fokker's sales to the USSR by relaxing visa requirements for Soviet tradespeople and offering (limited) credit guarantees for exports to Moscow.[37] With the Soviet deals, the Nederlandsche Vliegtuigenfabriek rapidly grew into one of Holland's major export industries. Fokker publicly stated that of the 196 aircraft sold abroad in 1924, with a turnover of 3,949,842 guilders, 176 went to the Soviet Union.[38] How these aircraft actually left the country nobody seemed to know. Few exports showed up in Dutch customs records: in 1924 Fokker reported not a single airplane shipped between January and August. The Netherlands Finance Ministry only learned of the Soviet sales informally, through circles of Russian emigrants in Paris and Vienna in June 1924,[39] prompting questions from the Justice Department. One of the country's leading newspapers also got wind of the affair. For a neutral and devoutly Christian nation like Holland, which refused to recognize the "pagan regime" in Moscow, major arms sales to the Soviets were a serious embarrassment. Sjirinkin, who had remained in Holland for a year to monitor the construction of the aircraft, hastily left the country, after which the affair was quickly hushed up.[40]

The confusion surrounding the Soviet patronage was greater because it coincided with secret German orders in the clandestine German rearmament program. Fairly soon after the signing of the Armistice Agreement between the Allied and Associated Powers and the defeated German Empire, members of the German General Staff started thinking about the possibilities for a future resurgence of Germany's military power. From 1919 on, the new army's chief of staff, Gen. Hans von Seeckt, played a crucial role in this planning. Von Seeckt worked untiringly toward some form of cooperation between Germany and the Soviet Union, which he intended both as a counterweight against the Allied victors of the war and a hedge against too strict an adherence to the provisions of the Versailles Treaty. The demilitarization of Germany, as demanded in that treaty, allowed for an army of no more than a 100,000, to be kept by the Weimar Republic only as a reserve force in the event of internal unrest. This requirement was simply unacceptable to the diehards who had comprised the upper echelons of the imperial army. Von Seeckt, and others, therefore pursued various ways in which Germany's lost position as a major European power might be regained. The initial solution, at least, was found in the East: the Soviet Union.

Since the October Revolution of 1917, the Bolsheviks, led by Lenin and Trotsky, had secured power in Russia for the Communists with amazing speed. Their ruthless drive led to the execution of the czar and his family, the unilateral annulment of all foreign debt, and the nationalization of foreign assets and interests in the country. By doing so, the newly formed Soviet regime stigmatized itself as the pariah of the international community of states. And that was only the first step. Within a week of the signing of the German-Soviet peace treaty of Brest-Litovsk on March 3, 1918, an Allied invasion force landed on Russian soil at Murmansk, marking the start of a war of intervention that lasted until 1921. The newly established order in the Soviet Union swayed. Army units refused to recognize their new Communist masters and rebelled. Together with the Allied intervention forces, these Whites, as the rebellious units were known, threatened to surround and overrun the young Soviet republic. The leader of the newborn Poland, Marshall Józef Piłsudski, cast hungry glances at the granaries of the Ukraine. In May 1920 the Poles joined the fighting but were unable to consolidate their initial gains against the Red Army. By August Soviet troops stood within firing range of Warsaw and could only be repelled with French support (Britain and France were the guarantors of Poland's independence).

The Polish-Russian war drove the international outcasts, Moscow and Berlin, closer together. Germany was the first country to recognize the Soviet Union, on May 7, 1919. Slowly and secretly, a process of German-Soviet negotiations got under way, which, on the German side, was conducted simultaneously on three levels: the government, the army, and private firms. They involved Von Seeckt, captains of the war industry such as Hugo Stinnes, the Krupp firm, and representatives of the Soviet government in Moscow. In these negotiations the Germans aimed to relocate parts of the German war industry on Soviet soil and thus evade the terms of the Versailles Treaty. The underlying premise of the whole process was that in exchange for a share in the production of such German factories, and German assistance in expanding and training the Red Army, Moscow might be willing to allow Germany to maintain some (secret) military facilities in the Soviet Union. Thus the clauses of the Versailles Treaty on the size and purpose of the new German army could be circumvented.

In the Kremlin, Lenin and his supporters showed themselves to be receptive to these ideas, provided they were part of a general treaty of assistance between the two countries. In the course of the negotiations, several German-Soviet joint ventures were founded, such as the Baltic shipping line Derutra (Deutsch-Russische Transport Unternehmen); a metal-trading company, Derumetall; and an airline, Deruluft (Deutsch-Russische Luftverkehrs Gesellschaft), which operated a scheduled air service between Königsberg in East Prussia and Moscow using German-built Fokker F.III aircraft.

On April 16, 1922, these developments culminated in the signing of the German-Soviet Treaty at the general European trade conference in Rapallo, near Genoa, Italy. In the weeks before that conference, diplomats of the two countries had already reached agreement on the main points of the treaty: German diplomatic recognition of the Soviet Union was to become public, and bilateral and trade relations were to be fully reinstated. This, in turn, opened the way for further talks between Von Seeckt and the Soviet General Staff on the sort of military cooperation that the Germans sought. Finally, a top-secret accord was reached in August 1922 allowing Germany three military training facilities inside the Soviet Union: a base in Saratov for research into poison gas warfare; one in Kazan for tank exercises; and an air force base near Lipetzk, a Russian resort town on the Voronezh River, three hundred miles south of Moscow. In exchange, the Soviets were to receive annual remuneration, German technological know-how, and various forms of military training.[41]

Fokker's exports to the Soviet Union were also connected to the German-Soviet agreement of August 1922 and the international political climate of 1923. The German order came about in the aftermath of the French occupation of the Ruhr in January, when the German president, Friedrich Ebert, requested that one hundred fighter aircraft be acquired abroad for future defense of German territory.[42] It was time for Fokker to pay off his debt to Germany for allowing him to railroad his aircraft and factory inventory to Holland in March 1919. In secret, lest the British or French secret service find out and the Netherlands government be warned to stop the exports, the Germans ordered an initial batch of fifty Fokker D.XI fighters, and a further fifty improved-performance D.XIII fighter aircraft for 3.5 million guilders. Fokker had no qualms about bending the rules and welcomed the lucrative scheme when visited by an obviously very secretive delegation of ex-Idflieg officers in his home in Amsterdam.[43] To confuse nosy individuals about the true nature of these sizable orders for first-rate military aircraft, the sales contract was drawn up via an intermediary: the Traugott Müller subsidiary of the German Stinnes

Concern. Labeled for export to Brazil, the fifty D.XIII's were shipped to Lipetzk, while the D.XI's, which the Germans decided they would not need after all (even though they had already been completed), were resold by Stinnes. Forty-eight of them ended up in the service of the Romanian Air Force.[44]

If anything, the scale of the sales to the Soviet Union (including the Stinnes deal) showed that the Dutch government had little idea of what Fokker was doing—and that the authorities were not particularly keen to find out, either. Though closer tabs were kept on Fokker exports for a while after the news of the Soviet sales had broken, official inspectors found there was really nothing they could do but report the *official* destination of the aircraft, as given in the accompanying paperwork. Whether or not these were indeed the final (or true) destinations remained unknown. Export papers stated Austria as the final destination of the forty-eight Fokker D.XI's resold to Romania when they left Holland by rail; the remaining two D.XI's left aboard the steamship *Euterpe,* bound for Denmark. Custom officials admitted having no idea whether the aircraft would go on to the Soviet Union, or not.[45]

Not wishing to rock the boat too much, Fokker played it straight for a while after that. Negotiations with Poland on the possible sale of and/or the rights for license-production of 300 military aircraft were conducted through the Netherlands legation in Warsaw, and sales to Spain and the Scandinavian countries were also duly reported.[46] In all, the Nederlandsche Vliegtuigenfabriek's foreign sales of military aircraft numbered at least seven hundred between 1919 and 1929. In 1924, the best year, 347 aircraft left the factory.[47] On top of that came foreign license-production of Fokker aircraft, which provided substantial extra earnings to the company. In 1926, for example, a contract worth eight million Czech crowns was signed with the Czechoslovak government for production of various Fokker types at the Skoda Works.[48] Likewise, numerous Fokker aircraft were license-constructed in Great Britain by the A. V. Roe Company,[49] in Italy by Romeo, in Hungary by Manfred Weiss, in Japan by Mitsubishi, and in Denmark, Norway, Sweden, and Switzerland.

In the context of such large exports and licensed production abroad, deliveries to the Netherlands Air Corps were of only secondary importance. Still, between 1919 and 1940, 368 aircraft were delivered to the Air Corps, 215 of them before 1930.[50] In addition, Fokker built some 90 airplanes for the Dutch Naval Air Arm and 87 for the air force branch of the East Indies Army, contracts that were the result of an understanding reached in 1920 after Fokker had strongly urged the Netherlands Ministry of War to provide the means that would enable him to keep his factory in Holland: "We wish to request Your Excellency . . . for support, not in the form of a subsidy, but in the form of a pledge that Your department will place its orders with us, if at all possible, and will also buy such prototypes as are necessary for a proper development of aircraft, when they are found to be acceptable to the State."[51]

## FIXING THE BOOKS

If the Dutch government considered itself indebted to Fokker for securing the neutral country's autarky in aircraft production—indeed, to the extent of granting him a virtual monopoly on its military contracts—the appreciation was not always mutual. Fokker found it difficult to operate within the strict laws and regulations of Dutch society. Bending the rules was something of a game to this nonconformist, self-made man. To him, it was fun, sporty, to defy the stuffy self-satisfaction of the various Dutch companies and government agencies he dealt with.

Most of such incidents went undocumented, except for the case of the performance trials held by the air force branch of the East Indies Army to decide finally whether or not to buy the Fokker C.IVB light reconnaissance-bomber, essentially an enlarged and re-engined C.I. The trials took place in the summer of 1923, after the standard C.IV had been in production for a year. So Fokker knew quite well what it could and could not do—and so might the East Indies military also have known, for the type was flown by the Dutch Air Corps. Nevertheless, the colonial army had drawn up a specification for the airplane that Fokker knew it could not meet. On the basis of various promises, he had secured an order for an initial batch of ten, anyway. The only catch was that the prototype of the "improved version" for the East Indies would be evaluated before series production began.

Fokker knew quite well whom among his factory staff he could call upon to "guide" such trials, and a team of "specialists" from the Fokker factory was present at all times to ensure unhampered testing of the aircraft. Those "specialists" worked extraordinarily long hours, well beyond those in which the military authorities were looking over their shoulders.

Though it took several years for some of the story to come out, in 1925 colonial authorities heard from an ex-Fokker employee who had decided to clear his conscience.

The correspondence between alarmed officials concerning this matter gives rare insight into Fokker's approach to business. Barographs, used to record the maximum attained altitude of the aircraft, had been rigged by Fokker's men to show the specified results. When the testing commission, perhaps knowing something about the way Fokker did business, had the instruments sealed after the flight for recalibration the next day, Fokker saw to it that their car was declared unserviceable. Schiphol Airport, the testing site, was some distance from Amsterdam, so Fokker arranged for the commissioners to spend the night in the small wooden airport hotel, next to the waterway connecting Schiphol to the Fokker factory. Next morning, a boat from the Fokker company showed up to take the party to Amsterdam. The boat took plenty of time for the trip. The committee members were wined and dined lavishly, and they got slightly drunk in the process. In the meantime, Fokker's accomplices got hold of the barographs, carefully removed the seals, and adjusted them to show their proper readings when recalibrated later that day.

Speed trials were similarly rigged. To mark off 1,000 yards on the airfield, at the beginning and end of which the aircraft would be clocked, the committee used two long measuring tapes. Fokker made sure that one of these tapes had several yards missing in the middle—not enough for the commission to notice, but enough to shorten the distance by about 50 yards. Obviously, the clocked speeds of the prototype aircraft turned out to be higher than those achieved by the subsequent production models.

And that was not all. For purposes of comparing other airplanes in trials, the contract specified airscrews 2 meters (6.56 feet) long. The Fokker prototype had an airscrew of 2.05 meters, although the length indicated on the propeller was 2 meters. Nor was the engine always quite what it was purported to be. In yet another test, Fokker switched engines in the middle of the night, and had the serial number of the original engine, which the commission had recorded, carefully welded to the engine block of the replacement. Contracts also specified a certain rate of climb to a predetermined altitude, taking off with a full fuel tank. In one trial, the Fokker mechanic, who accompanied the pilot, was told to pull a little wire that had been attached to a hidden relief valve in the fuel tank to reduce the weight of the plane, knowing the fuel level in the tank would not be checked again upon landing.[52] Learning the extent of Fokker's tricks, the

Rijks Studiedienst voor Luchtvaart (Government Study Agency for Aeronautics) immediately set about amending procedures and instrumentation for such trials.[53]

## GLIDING ALONG

How closely Anthony Fokker watched the exploits of his Dutch company is unclear. From 1922 on he was more and more often away from Amsterdam, pursuing new markets in the United States (see chapter 6) or pursuing the pleasures of life available to an *homme arrivé*. Ever so slowly, his life was starting to take on a more relaxed form, and, combining his flying skill with the kind of gratification he had derived from experimenting with his early machines, he developed a strong interest in gliding.

That interest dated from mid-August 1921, when Fokker and Bernard de Waal drove Tony's big eight-cylinder Cadillac from Amsterdam to the tiny German ski resort of Gersfeld, 120 miles east of Frankfurt, at the foot of one of peaks of the Rhön Mountains, the even-sloped Wasserkuppe. They went there hoping to recapture some sense of their bygone Johannisthal days, visiting the second annual Rhöner Segelflug Wettbewerb (Rhön Gliding Contest). Forty contestants entered the competition, a number of them old familiar faces from wartime Germany. They gathered as the outcome of the postwar Allied ban on private flying with powered aircraft in Germany, which had stimulated the novel sport of gliding. Such activities usually took place under the auspices of student gliding clubs, which were frequented by war veterans.

The gliding scene was not quite as harmless as it appeared at first. Fokker noted the presence of ranking government officials and that the army was taking a keen interest in what was going on in Gersfeld. Military interest was evident in permanent sheds, erected on the slope of the Wasserkuppe, and a number of tents for not-so-affluent contestants. But gliding was mainly for sport—the object was to hover over the slope of the Wasserkuppe as long as possible in a motorless airplane.[54] Fokker returned to Amsterdam full of enthusiasm: this was *real* flying again, the way it had been before the war. Gliding had the old excitement of danger, because relatively little was known then about thermals and how to use them for unpowered flight. At the Septième Exposition Internationale de Locomotion Aérienne, the most prestigious international aeronautical fair, held in Paris between November 12 and 27, 1921, the Nederlandsche Vliegtuigenfabriek exhibit reflected Fokker's new hobby. Two airplanes were on display: Fokker's latest model F.III airliner, chartered from KLM, and an engineless derivative of his wartime D.VIII fighter, the similarities of which did not entirely escape the French audience.

Insisting to attend the Paris exposition in person, Fokker caused quite a stir there. First, both of his airplanes carried the customary "Fokker" logo on the fuselage, which had to be painted over hurriedly after what the French paper *Le Matin* called the "outraged protests of old combatants, and mutilated victims of the war."[55] The next day, the French under secretary for Air, Laurent Eynac, himself an ex–fighter pilot, visited the Salon. When he approached the Fokker stand, the secretary of the Royal Netherlands Aeronautical Society, Isaac "Jumbo" van den Berch van Heemstede, who had been talking with Fokker, stepped forward and announced in a loud voice, "Monsieur Under-Secretary, I have the pleasure of presenting you the famous constructor monsieur Fokker!"[56] The result was pandemonium in the French delegation, from the midst of which someone shouted, "À bas les Boches!" (Down with the Krauts!). The French paper *Le Journal,* which had its reporter on the scene, reported that Fokker came forward to meet Eynac, fully expecting to shake hands. "His expression frozen, his teeth clenched, he [Eynac] made an effort, plain to see for all, to avoid the outstretched hand, and with an abrupt turn, left the stand."[57] Later, it was discreetly pointed out to Fokker that it might best serve his company's interests, and Franco-Dutch relations, if he were, in the future, to refrain from attending the Salon.

The hostile reception did not, however, diminish Tony Fokker's newfound romance with gliding. Late in July 1922, returning to Holland after three months in the United States, he immediately instructed Platz to build two gliders to take with him to the Wasserkuppe in a couple of weeks. Bending their backs as they had in the old days, Fokker, Platz, and a number of workmen at Fokker's factory subsidiary at the Veere Naval Yard built two road-transportable gliders in record time. Fokker reported in Germany on August 24 with the FG.1 and the FG.2, both light biplanes, a single-seater and a two-seater, inspired by the Wright designs of 1901–3 and with similarly styled wing-warping controls. Arriving at the Wasserkuppe with a high-powered crew of mechanics, equipment, and two trucks, Fokker was greeted with mixed feelings. He was clearly out of tune with the average German contestants, some of whom had spent their last pfennigs to participate. Besides, since the previous year, the frontiers of sustained, unpowered flight had been pushed forward from minutes in the air to several hours. Fokker therefore decided not to participate in the endurance trials, but his two-seat glider was something new. With Seekatz on board as his passenger, he established a world endurance record for two-seat gliders on August 26, 1922: 12 minutes and 53 seconds. Later that year, in October, Fokker also participated in the glider meet at Itford Hill in Sussex, England, achieving a respectably long flight of 37 minutes and 6 seconds on October 16. After that his short career in gliding came to a abrupt stop.[58] Considering the participants at the Wasserkuppe and Itford Hill, he decided that his approach to gliding set him apart from the other contestants, who did not have his affluence and a technical staff behind them. Gliding might be fun, but the people involved did not create the sort of environment the millionaire aircraft manufacturer felt quite at home in. Fokker would never again board a glider.


## PRICES TO PAY

Fokker's German past also influenced the way he organized the Nederlandsche Vliegtuigenfabriek. Throughout the interwar period, Fokker welcomed license-production of his airplanes, since his factory had problems handling sizable orders. Striking a balance between investment in machinery and investment in work force was an omnipresent issue that had shadowed Fokker's policy ever since the sudden influx of German army orders in the months leading up to the outbreak of the First World War. In the postwar period, this was a constant dilemma, too, and one that Fokker did not care to resolve. Constantly bickering with his managers, Fokker put off long-term investments, even if they were vital to the company.[59] His policy was one of contracting out large components to meet immediate demands exceeding work-force production capacity, not one of rationalizing production. For example, wooden wings for the aircraft exported to the Soviet Union were constructed by Werkspoor, the railcar-production division of the Dutch national railway company.[60] To complement subcontracting, Fokker adhered to a policy of drastic hiring and firing, which earned him a bad name. By the summer of 1920, his factory had 147 people on its payroll, a number that rose to about 1,300 in 1924 and then dwindled rapidly to 385 at the beginning of 1926, when orders dried up.[61] Production stagnated, and Fokker's new deputy director, Bruno Stephan, whom he hired in February 1925 to look after his business in Holland while Fokker himself went to the United States, was faced with the unpleasant task of organizing mass layoffs from the moment he first walked through the factory gates.[62] When the market improved toward the end of the 1920s, Fokker again resorted to taking on more employees (595 in 1929) to meet current demands rather than committing himself to long-term investments in modernization of the factory.

Nevertheless, Fokker's Nederlandsche Vliegtuigenfabriek did reasonably well, considering its generally disorganized state of affairs—to the credit of the people Fokker hired to run it for him. The (net) profit of the company as related to total turnover (which included substantial amounts of money flowing from licensed production) averaged 7 percent annually between 1925 and 1935, or 1,603,500 guilders. On top of that, Fokker received a total of 1,645,000 guilders from license rights in the same period.[63]

Yet preciously little was done with these profits. Without sufficient reinvestment and capital expansion as any business grows, it will run into difficulties in the long run. Fokker's business was no exception. Instead of investing some of the company's reserves in factory improvements (fixed assets) or fundamental research, Fokker preferred to muddle on in the ELTA buildings. By 1929 they had been completely written off on the company's balance sheets. Cash reserves already amounted to 3.5 million guilders, while capitalization and shareholding remained unchanged.[64] Repeated offers from Frits Fentener van Vlissingen, Fokker's prominent Board of Control member and corporate investor, to issue additional shares to expand the company's basic strength were rejected by Fokker time and time again for fear of losing control over *his* company.[65] Fokker saw profits as the ultimate proof of the soundness of his business practices: clearly, factory modernization and investment in radical product innovation were unnecessary. All was going well—or was it?

Throughout the 1920s, Fokker's investments in research and development, inadequate though they may have been, were heavily tilted toward military aircraft. Here, he stood the best chance of recouping investments through big export orders. Indeed, Fokker's military sales in the 1920s outnumbered his civil sales by four to one. Such military exports were mainly a result of the fame that the fighting Fokkers had acquired in World War I, whereas the reputation of Fokker's postwar products had to be carried far and wide by his civil types from the mid-1920s onwards (see chapter 5). Despite relative underinvestment in civil aircraft development, here was the technological core of the company's achievements.

Fokker failed to come to terms with these realities, even though he was quick to pick up on the publicity that the success of his commercial airliners brought. But concentrating investments in research for and development of military aircraft—in the haphazard and thus costly way in which Fokker and Platz preferred to work—meant that the real challenges of technological improvement, which should have accompanied the dramatic increases in size of Fokker airliners in the second half of the 1920s, were evaded. After an innovative but unsuccessful low-wing monoplane fighter, the D.XIV (developed in 1925) proved costly and, even so, unsatisfactory, new military and commercial aircraft designs alike became increasingly conservative in their approach.[66] This conservatism was a reflection of the prominent position that Platz had acquired, for Fokker's chief constructor had completely lost touch with technological developments in aircraft construction and found himself unable to accept new ones.

Though the Fokker company was not the only one to show the effects of shrinking military budgets on research and development, its lukewarm approach to innovations was bound to have consequences. At the cutting edge of European aircraft technology, such German constructors as Hugo Junkers, Claudius Dornier, and Adolf Rohrbach continued to invest in the development of metal aircraft, but Fokker failed to keep up.

In the late 1920s, the real challenge in aircraft construction came from the growing airline industry. The most innovative designers were taking big strides toward with more economical planes, built according to a scientific approach, rather than the pragmatic, rule-of-thumb methods that Fokker and other early producers continued to prefer.

In California, Malcolm and Alan Loughhead and Jack Northrop of the emerging Lockheed Aircraft Company took American aircraft from the rugged and proven types built by such constructors as Curtiss, Douglas, Boeing, Stout, and Ford to aerodynamically clean aircraft based on advanced design methods. Their eye-catching and remarkably well-streamlined Vega incorporated such novelties as a stressed (plywood) skin fuselage of semimonocoque construction and a much higher wing loading than was usual at the time. Its rapid commercial success—Lockheed sold 129 Vegas before production ended in 1934—made it a trendsetter in American aircraft design, which entered a phase of rapid progress.[67]

These developments were stimulated in 1930 by the adoption of the third amendment to the United States Air Mail Act of 1925, known as the McNary-Watres Bill, which contained incentives for operators and designers alike to produce larger and more economical transport aircraft.[68] Spending most of his time in the States, where, on December 14, 1923, he had founded a new company, the Atlantic Aircraft Corporation, Tony Fokker was well aware of such trends yet found it impossible to adjust. Despite the success of his F.VII series, Fokker kept aiming at the military (export) market that had traditionally made up the majority of his sales.

It was a fateful decision. The stock market crash of 1929 and the worldwide economic crisis that followed prompted severe cutbacks in military spending that expressed themselves through the League of Nations in an international drive for disarmament in the early 1930s. Fokker's traditional markets for military aircraft dried up.

But civil aviation was only mildly affected by the economic slump. Protected and financed by their respective national governments, the airlines were seen as manifestations of national prowess and therefore maintained for reasons of national prestige, at least as much as for commercial reasons. By continuing to concentrate on military aviation and failing to adapt to the trend of engineering innovations in airliner design, Fokker stood to lose both his military *and* his civilian market after 1929. Under these circumstances his policy of marginal investment was becoming a millstone around the company's neck. Even without the economic crisis of the 1930s, the Nederlandsche Vliegtuigenfabriek would have fallen into deep trouble. Fokker's own rule-of-thumb methods of working and the failure of Platz, his equally uneducated chief constructor, to keep tabs on aircraft development in Amsterdam brought their long association to an end. It left no room for changes that would take aircraft development beyond their personal comprehensions and capabilities. After considerable and protracted deliberations, Fokker finally resolved to let his old-time associate Platz go in May 1931.[69]

Another hour of education lost: Tony Fokker *(in the rear)* and his friend Frits Cremer row their boat across the river Spaarne in Haarlem, Holland, in 1909.

Fokker's early associate and flight instructor, Bernard de Waal (died 1924), poses in front of a 1912 Spider on his Dutch-built Vulkaan motorcycle.

Happy days: Fokker's "girlfriend from St. Petersburg," the vivacious Russian aviator Ljuba Galantschikova, poses for her Dutch admirer on top of a Spider in Johannisthal in 1912. When she left Fokker to go off with the French pilot Léon Letord, she broke his heart.

Fokker's friend and partner, Frits Cremer, whose father helped put Fokker in business, was a flight instructor at Fokker's flying school in Schwerin (late 1913).

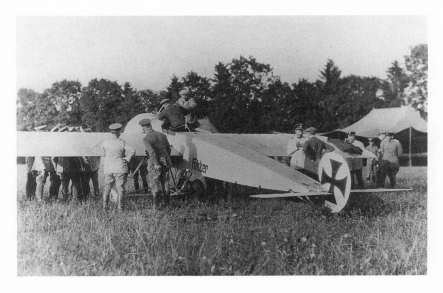

Demonstrating a revolution in air warfare: Tony Fokker *(behind the fuselage, with cap)* talks to the German crown prince, Wilhelm, at his headquarters in Stenay, northeastern France.

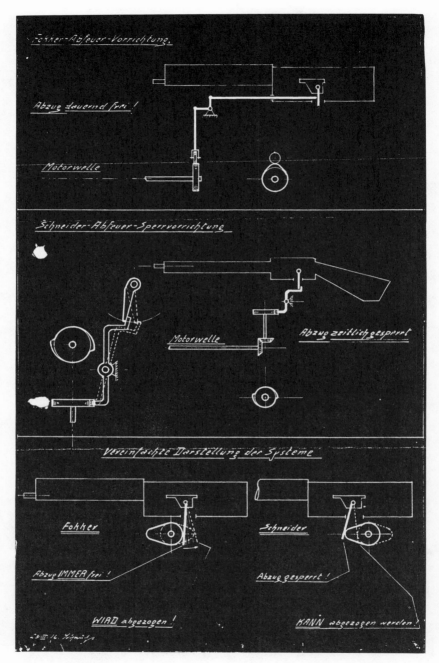

A blueprint showing the similarity between the Schneider and Fokker machine-gun synchronizing mechanisms (1915). The intermittent legal battle over patent rights to the invention continued for seventeen years.

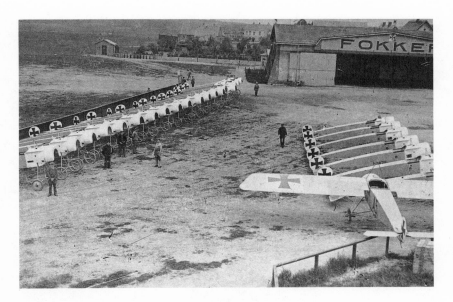

Engine production did not keep pace with aircraft production in wartime Germany: twenty-five Fokker E.III's await delivery of Oberursel rotary engines at Schwerin (1915).

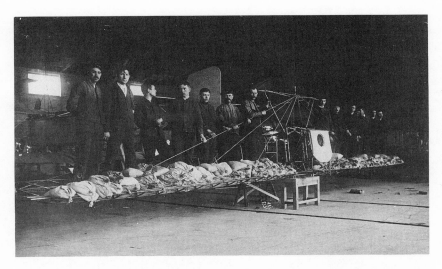

Schwerin employees and sandbags on an inverted wing of one of the E-type monoplanes, a public demonstration of stress testing, Fokker style.

A more closely monitored endurance stress test with sand. Elaborate testing did not prevent Fokker aircraft from suffering repeated wing fractures, resulting from poor quality control in series production.

Tony Fokker sits in the cockpit of the radically streamlined V.1 (early 1917). This aircraft was the first to incorporate the cantilever wing construction. Poor handling characteristics and downward visibility from the cockpit led to the abandonment of the project.

Tony Fokker *(second from left)* talks to German flying aces Kurt Wolff *(third from left)* and Manfred von Richthofen *(right)*, and Jagdgeschwader I's technical officer, Konstantin Kreft *(left)*, at Schwerin on the occasion of test-flying the new F.I/Dr.I triplane in early August 1917.

Tony Fokker *(extreme left)* shows the prototype of his F.I triplane to Von Richthofen *(just right of the tailplane)* and skeptical-looking army officials at Marckebeeke, Belgium (late August 1917).

Operational deployment of the Fokker Dr.I triplane took place in France early in 1918. On forward bases, it was not unusual for aircraft to be housed in tents *(background)*. Wet conditions at the front caused humidity problems in the wings.

Fokker, uncharacteristically all dressed up, poses in front of a Fokker V.12 fighter with Friedrich Seekatz, his head of aircraft production at MAG in Budapest, Hungary, in the spring of 1918.

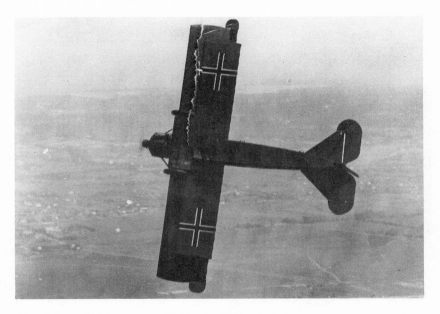

The ultimate fighter plane, a Fokker D.VII banks in a sharp turn over northern France in 1918. The success of this aircraft necessitated Fokker's move to Holland in February–March 1919.

A factory to be abandoned: Fokker's workshops at Schwerin-Görries, seen here in early 1918. Shunning investments in fixed assets, Tony Fokker preferred to house his expanding production in temporary wooden sheds.

A trainload of Fokker aircraft rolling west. Railroads moved aircraft to the front. Fokker also used them to transport over two hundred aircraft and a substantial part of his factory inventory across the border into Holland.

Although very publicity-minded when it came to business, Fokker shrouded his personal life in secrecy. Here Tony and Tetta emerge from Haarlem's town hall after their civil wedding ceremony on March 25, 1919, for which Tetta did not wear a bridal dress. (At that time in Holland, not to marry in church was a rare and slightly scandalous thing.) In the rear, Herman Fokker could hardly wait to light his cigar.

The happy aeronautical couple, Tony and Tetta actively sought the eye of the press to draw attention to Fokker aircraft. Here, Tetta is decked out with a bouquet of roses before a publicity flight from the military airport at Soesterberg to the Scheveningen beach (where Fokker wanted to start his own flying school) and Schiphol airfield, near Amsterdam, on May 30, 1919.

Low-profile big business in the early 1920s: a crated fighter aircraft destined for the Soviet Union is loaded on board the freighter *Europa* in the Amsterdam docks. No questions asked.

Trying to recapture the spirit of old times, Tony Fokker and a small entourage assemble the two-seat FG.2 glider at the Wasserkuppe gliding contest in the Rhön mountains in Germany (August 24, 1922).

Some of Fokker's unsold stock of aircraft brought from Germany slowly deteriorating in the open air at his Dutch factory, formerly the ELTA exhibition halls, in 1923.

An overcrowded workshop in the ELTA factory at the height of Fokker's production in 1924. Fokker followed a rigorous policy of hiring and firing to meet fluctuations in demand.

Fokker's first commercial airliner was the F.II of 1920. To prevent air sickness, the windows could slide open to let in fresh air. A female passenger in the back seat is leaning out, waving a scarf.

Royal patronage was important for Fokker's position in a small country like Holland. Fokker, in his best attire, is shaking hands with Queen Wilhelmina, while his friend, Prince Consort Hendrik *(extreme left),* and Crown Princess Juliana look on (April 3, 1924).

A Fokker F.VIIA, workhorse of Europe's emerging airlines, flies over a typically Dutch waterlogged landscape.

Waterskiing Fokker-style on a self-built contraption. The water was Fokker's favorite environment for relaxation.

On the slopes of Saint Moritz, Switzerland, Fokker posed with Oberalpina, his chalet and customary winter retreat, in the background.

Tony Fokker walks away disgruntled after another postponement of Richard Byrd's flight in *America*. He never forgave Byrd for not being the first to fly nonstop across the Atlantic.

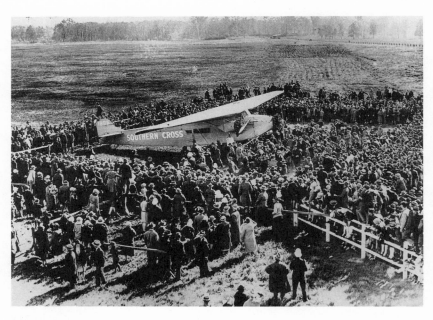

A more successful oceanic crossing: Kingsford-Smith's long-span F.VII trimotor *Southern Cross* landed at Eagle Farm near Brisbane, Australia, after making it across the Pacific in June 1928.

Fokker looks approvingly at his flying-billboard entry in the Ford Reliability
Tour. His ascent in American aviation began in October 1925 with the public
acclaim of his first trimotor.

Fokker's main production plant in the U.S. on the airfield of Teterboro, New
Jersey, just south of New York City. Several of Fokker's unsold F.32's are sitting
on the tarmac in this 1930 photograph.

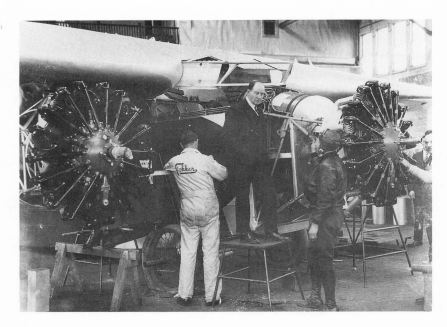

An exacting supervisor, Tony Fokker checks the workmanship of one of the F-10's under construction at the Teterboro factory.

Tony with his second wife, Violet, pose for newspaper photographers at the Chicago Expo (November 30, 1928). On February 8, 1929, their marriage was to end in tragedy.

Fokker *(center)* talking to his new General Motors appointee as vice president for sales, Eddie Rickenbacker *(left)*, and an old-time customer, the Arctic explorer Sir Hubert Wilkins (September 1929).

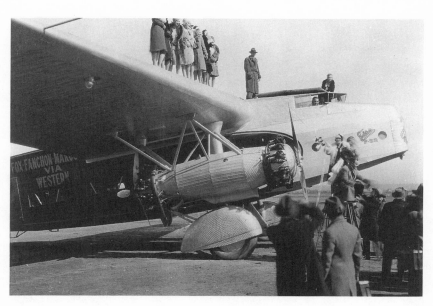

Cheerful festivities accompanied the acceptance of the first F-32 by Western Air Express on March 21, 1930, when Fokker still had hopes of marketing the giant airplane.

Product diversification—building a different plane for every niche in the market at the cost of fundamental research and development—lay at the bottom of the demise of the Fokker Corporation. *Left to right:* Universal, F-14, F-11, F-32, F-10A, and Super Universal.

At the scene of the notorious Rockne Crash near Bazaar, Kansas, on March 31, 1931. The disaster led to a temporary grounding of the F-10A's that destroyed the reputation of the Fokker Corporation. But the real reasons for the downward slide of the company lay deeper.

Wing construction for the F-10A at Fokker's new production plant in Wheeling, West Virginia (January 1929). Difficulties in maintaining quality control with an inexperienced work force led to the wing defects held responsible for the Rockne Crash.

Early NACA engine cowlings were fitted on a U.S. Army Fokker C-2A at Langley Aeronautical Laboratory in 1929. When initial results indicated that these aerodynamic refinements did not significantly increase the aircraft's speed, Fokker was encouraged to continue aircraft development according to his proven design.

Dutch successes with Fokker aircraft: a cheerful crowd waves goodbye as the first of a series of biweekly KLM trial flights to Batavia (Jakarta) takes off from Amsterdam's Schiphol Airport on September 12, 1929.

The ultimate low-budget airliner, five Fokker F.XVIII's were sold to KLM for its Amsterdam-Batavia route. Research and development for the aircraft cost less than $300.

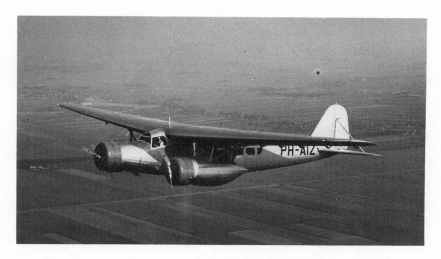

To put his American experience to work while continuing to control costs, Fokker developed the F.XX airliner in 1931–32, only to find that his habitual launching customer, KLM, refused to back the plane.

Tony Fokker talks to Dutch aeronautical officials under the giant, prohibitively expensive F.XXXVI airliner. In the background, the D.XVII fighter of the same period epitomizes the deep technological rift between civil and military aircraft before the mid-1930s.

The F.XXXVI incorporated an unusual cockpit layout, which, Fokker claimed, vastly improved the pilot's vision. Here he is seen explaining the details to Charles Lindbergh *(left),* with KLM's Albert Plesman watching skeptically.

Outclassed technologically, but not commercially, Fokker points the way ahead for himself and his company by selling Douglas airliners in Europe (1935).

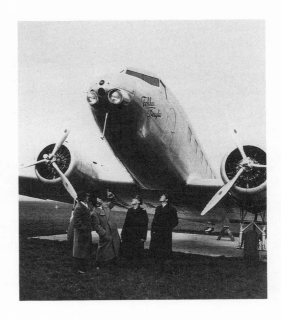

Boating and floating: Fokker cavorts with two female guests near his favorite ship, the *Honey Moon,* anchored in the background.

Fokker and Swissair manager Walter Mittelholtzer strike an informal pose on the frozen lake in Saint Moritz, Switzerland, in front of a DC-3. Fokker's sales activities on behalf of Douglas and Lockheed were vital in establishing the hegemony of the American aircraft industry in the market for commercial airliners.

Fokker combined aeronautical and marine technology in designing his luxurious yacht, the *Q.E.D.* He considered the $200,000 ship his biggest achievement.

Pleasure in the eye of the beholder: Tony Fokker contemplates female beauty late in life.

# 5

## TO CAPTURE A WORLD MARKET

In July 1920, a year after the founding of his Dutch enterprise, Tony Fokker still found himself uncomfortably close to the sidelines in the field of postwar aviation. His new company was not making much money; since the Trompenburg contract, no new business breakthrough had come his way; his initial approaches to KLM had been rebuffed; the Netherlands Postal Service had not accepted his airmail bid. And with an abundance of surplus military aircraft on the international market, foreign sales were not altogether likely to materialize either. Fokker clearly saw that to improve his foothold in postwar aviation he had to secure some new orders soon.

In the civil field, he pinned his hopes on the F.II, the new airplane that had been test-flown by Platz's crew in Schwerin. As airliners went, the design was little short of revolutionary. At a time when most airplanes were of a biplane configuration, and passenger-carrying types needed multiple engines, the F.II elaborated on a single engine of 185–240 hp and the single, thick, plywood-covered cantilever wing that had been developed on the basis of the Junkers patents and Forssman's wartime ideas. Suspended underneath the wing was an angular fuselage of welded steel tubes covered with doped linen and wood. Behind an open cockpit, the fuselage housed an enclosed cabin capable of carrying four passengers in cramped seating. It was the first of the Fokker series of transport aircraft to bring the company fame and fortune in the 1920s.

Fokker had good reason to believe that the Dutch national air carrier, Koninklijke Luchtvaart Maatschappij voor Nederland en Koloniën (KLM Royal Dutch Airlines) was the ideal launching customer for this venture into commercial transport aircraft. Selling his aircraft to the national airline would emphasize the Dutch character of Fokker's Nederlandsche Vliegtuigenfabriek. Fokker needed that national element, and Holland's neutrality, to be acceptable on the postwar international market. The foreign airlines, costly expressions of national prestige, were still few in number but attracted high-level attention, including financial interest in Europe.

KLM had been founded by the major Dutch banks, financiers, and shipping lines on October 7, 1919. By July 1920, following French, Belgian, and German examples, and afraid of being left out of international air transport, the government had already decided that a Dutch airline was a valuable national asset, one that was worth subsidizing. Besides, having a national air carrier held up the prospect of future air communications with colonial territories in the Dutch East Indies.[1] By the same token, KLM was the logical launching customer: the Netherlands airline should operate Dutch-built aircraft. Moreover, because KLM did not yet have any aircraft of its own, operating a single air service (to London) under a provisional charter contract with aircraft of the British airline, Air Transport & Travel Ltd., Fokker faced no established competitor in aircraft deliveries.

Hence Fokker went out of his way to persuade KLM to buy his new airplane. Besides praising its technical characteristics and reciting nationalistic rhetoric, he offered the F.II for the rock-bottom price of 27,500 guilders to make the deal even more attractive.[2] The comparable British de Havilland DH-18, a single-engined four passenger biplane, cost more than two and a half times that amount.[3] With such an extraordinarily generous offer, Fokker had little trouble landing an order from KLM for two F.II's.

What Fokker did *not* mention in any of the discussions with KLM's management was that although he presented the F.II as a product of his Nederlandsche Vliegtuigenfabriek, it had actually been built by SIW in Schwerin. KLM's young managing director, Albert Plesman, found out about this only *after* the sales contract for the F.II transports had been signed. Both he and his Board of Directors were deeply distressed, accusing Fokker of double-crossing the airline. After all, the situation could create serious trouble for KLM: what if Holland's neighboring countries refused to grant KLM permission to operate commercial services because it used German aircraft? With resentment against all things German still running high, the aircraft might even be impounded upon landing in Britain, Belgium, or France!

That was if they got there at all, for the slightly illegal delivery flight of the first F.II from Schwerin in the spring of 1920 was marred by engine problems. After repeated emergency landings, the pilot, Bernard de Waal, literally ditched the airplane in a little watercourse surrounding a pasture near Surhuisterveen in northern Holland. The F.II reached Amsterdam by barge.[4] After the necessary repairs, it was delivered to KLM on August 25, 1920. To resolve the airline's anxiety, the F.II was then flown to London on September 30—a memorable "flight" in which the aircraft kept developing mechanical troubles and the pilot was forced to return to Schiphol no less than five times. Two days went by before all malfunctions were finally sorted out and the aircraft reached London.[5] To KLM's relief, the British did not impound its German-built airplane.

The apprehension about the F.II contract was the first sign of a lengthening shadow that darkened relations between the Nederlandsche Vliegtuigenfabriek and KLM—and in particular between the leading personalities involved, Tony Fokker and Albert Plesman. The privileged world of the eccentric millionaire and rich man's son, Fokker, and the humble background of the seventh child of small shopowners, Plesman, were mutually exclusive. Plesman had been chosen to head KLM on the basis of his exceptional drive as a young officer in the army and his untiring efforts to organize the ELTA. He was something of a bully, strong-mouthed, and every inch the soldier he had worked so hard to become. He could not help feeling a personal resentment against Fokker's upper-class background and the slick business practices that Fokker tended to use against him.[6] For Fokker's part, while pleased with the publicity the "all-Dutch" Fokker-KLM combination brought, he failed to come to terms with the chauvinistic connotations that underlay KLM's patronage. Heavily dependent on government subsidy, KLM had no realistic alternative to buying his machines, yet Plesman expected *more* than simply value from Fokker. Conflicts over prices and payments were a continuous source of friction: Fokker behaved whimsically in his contacts with KLM, tending to back away from earlier understandings. Besides that, quality control remained a constant problem for the Nederlandsche Vliegtuigenfabriek.

The difficulties over acquisition of the F.II were still fresh when the next controversy arose. Early in October 1920, with the two F.II's standing idle on the grass of Schiphol Airport, KLM was bound by its charter agreement with Air Transport and Travel until the end of the "flying season" on November 1. Fokker told Plesman that work was already under way on an improved design. At roughly the same all-up (gross) weight as the F.II, and powered by a similar engine, the F.III was to incorporate a wing of increased span (57.9 feet—1.2 feet longer than the F.II's) to provide more lift, and a shorter but wider fuselage seating five passengers. Plesman immediately showed interest, stating that of the two airplanes, he would much prefer to put the F.III into service when KLM resumed flying in March 1921. Fokker pointed out this was impossible because the airplane was only in its early development stages and would probably not be ready for airline service by then. Nevertheless, he urged Plesman to hurry up and place an order if he wanted to be the first to put this new, and more economical plane into airline service. Fokker was about to leave on a six-week trip to the United States and would not be able to guarantee KLM a position as the launching customer unless a firm order was placed before his departure. In that case, Fokker offered to deliver the aircraft at a price substantially below the one he would charge to foreign customers.[7]

Under financial pressure from KLM's shareholders, and involved in subsidy negotiations with the Dutch government stipulating that KLM's private investors should cover at least a third of the airline's expected short-term losses until the airline became profitable in a few years, Plesman received a go-ahead from his board to reach an agreement with Fokker as soon as possible on this more economical airplane. Such time pressures led to a serious misunderstanding that clouded relationships between the principals. Eager to obtain KLM's sponsorship and so reduce the financial risks of developing the F.III, Fokker suggested that he might buy back KLM's F.II's at invoice price if KLM decided to buy the F.III. Believing this to be a firm promise, a very pleased Plesman thereupon reached a verbal understanding with Fokker to buy two F.III's. But while Fokker was still in the States in December, his Dutch representative, Hendrik Nieuwenhuis, told Plesman that he knew of no such agreement. Returning in January 1921, Fokker took the same position, denying any obligations beyond KLM's order for two F.III's. There was no question of taking back the earlier aircraft, despite Plesman's demands to exchange them.[8] Fokker insisted on delivering the machines, and after an exchange of angry letters, and a nasty, high-pitched scene between the two men in Fokker's house in Amsterdam's Roemer Visscherstraat, KLM found itself with no option but to buy the F.III's.[9] But that was not the end of the conflict: after the aircraft had been delivered, an infuriated Plesman reported to his board, "The Nederlandsche Vliegtuigenfabriek tries to deliver bad materials, such as old tires, bad airscrews, bad window panes, etc."[10]

Meanwhile, the development of the F.III had taken a curious course, typical of the haphazard way aircraft development proceeded under the direction of Fokker and Platz. Improving on the design of the F.II, Fokker had personally proposed changes in the fuselage of the airplane in October 1920: it should be widened to accommodate three-abreast seating and made shorter. The pilot's seat was to be moved forward right beside the engine, thus offering improved, all-around vision and allowing for a bigger passenger cabin.[11] These changes were incorporated in provisional construction drawings made in Amsterdam while Fokker was away on his American trip. A crude trial model constructed there with this new arrangement became the basis for a more precise blueprint of the new airplane. This blueprint was sent to Schwerin, where the prototype was constructed accordingly—it being cheaper and more efficient to do so in the German factory. After completion, the prototype was dismantled, crated, and sent to Holland by rail to be reassembled in Fokker's test hangar at Schiphol Airport in November. On November 20, a test pilot from Fokker and KLM's senior pilot both made a short test flight in the new airplane and found it to behave satisfactorily. But because Tony Fokker, still his own ultimate test pilot, was in the United States, further trials were suspended until his return.

When Fokker came back in mid-January 1921, it took him just one flight to reject various parts of the fuselage construction, engine mounting, and control surfaces. This showed Fokker at his best, suggesting changes on the basis of practical, in-flight experience. Studying these changes, the Dutch airworthiness authority, Rijks Studiedienst voor Luchtvaart (RSL), later concluded that they made the F.III safer and easier to fly for less experienced pilots. After alterations, Fokker made further test flights, from which more modifications followed. New blueprints were produced and sent back to Schwerin for series production.[12] Before the end of aircraft construction activities at the SIW, in July 1921, Fokker imported ten F.III's from his German factory.[13] After that, German license-manufacturing of the F.II and F.III continued. The Grulich

firm built eighteen F.II's and eight F.III's, which ended up in the service of
the German airline Deutscher Aero Lloyd and the Soviet-German airline
Deruluft. Many of these aircraft later passed to Deutsche Luft Hansa
(DLH) when all the smaller German airline companies were amalga-
mated.[14] In all, some fifty F.III's were built.

Back in Holland, relations between Fokker and KLM continued their
downward slide. When the government indicated the airline should expand its private capitalization, Plesman was ordered by his board to talk
to Fokker about the latter's possible participation in KLM's shareholding.
To Plesman's dismay, Fokker immediately offered to invest an initial
35,000 guilders, provided he was given a seat on the Board of Directors.
Plesman had to use much persuasion to convince his shareholders that accepting Fokker's bid was a bad idea, one that would wipe out all of
KLM's already narrow leeway in aircraft procurement.[15] The soundness
of this position would soon become evident.

With the intended expansion of KLM's services, the airline had become
interested in acquiring six additional aircraft. Fokker proposed his F.V, a
biplane that could be reconfigured as a parasol-winged monoplane, trading payload capacity for higher speed. He offered the F.V at a price of
45,000 guilders each. In April 1923 Plesman and Fokker's Dutch company lawyer, J. H. Carp, one of the few people to whom Fokker trusted
(limited) powers of attorney, held initial talks on the issue. Wary of
Fokker after their earlier dealings, Plesman tried his best to ensure top-
quality aircraft at the lowest possible price, stating that the F.V was still
too expensive for KLM. To put pressure on Fokker, he encouraged
Fokker's tiny "competitor," the Nationale Vliegtuig Industrie (NVI: National Aircraft Industry), a small-scale, underfunded venture in aircraft
construction that employed Frits Koolhoven as its principal designer, to
come up with a competing design.

Fokker thereupon lowered the basic price of his airplane to 40,000
guilders, still more than the 36,000 KLM had in mind. Drawn-out negotiations followed, until, when discussing the matter with Plesman, Fokker
suddenly made a very generous offer on May 25, 1923: he would allow
KLM to test his F.V in regular airline service to establish its suitability
and proceed at the same time with the development of another model, the
F.VII, giving KLM an option to buy three to six of either model for
36,000 guilders each. Work would be simultaneously undertaken on
twin-engined and three-engined variants of the F.VII, which Fokker offered to KLM at a price of 100,000 guilders for three (excluding the

engines).[16] Such generous terms could hardly be passed up, and Plesman was very pleased to accept Fokker's offer. Not four months later, when KLM was already committed to proceed with Fokker aircraft, Fokker dropped a bombshell. In a letter to KLM of September 13, he took the position that no legally binding agreement had been reached concerning the order; indeed, nothing about his generous offer had been put in writing. Fokker now proposed to draw up a formal sales contract for the single-engined F.VII at a unit price of 61,000 guilders! KLM's board was outraged at such a breach of good faith. While instructing Plesman to keep discussions going with Fokker on a lower price for the F.VII, just in case, KLM's executives decided to look for ways to cancel the order without impairing the airline's projected expansion of services.[17] After another month of heated discussions, Fokker finally gave in and allowed KLM to buy three F.VII's at the bargain price of 36,000 guilders each, as had originally been agreed upon.[18]

By that time, the prototype of the F.VII, a high-wing monoplane with an enclosed cabin that now seated up to eight passengers and was powered by a single 360 hp Rolls Royce Eagle engine, was nearing completion in Fokker's Amsterdam factory. The aircraft was conceived and built under supervision of Platz's adversary within the Fokker organization, the German chief designer, Walter Rethel, who had joined Fokker from the Kondor Werke in 1919. On January 17, 1924, the aircraft arrived at Schiphol by barge for final assembly. However, it was soon discovered that a clumsy and fundamental error had been made in the design and construction of the aircraft: when the propeller was mounted, its axle was a foot (30 cm) too low, so that the propeller would hit the ground as soon as the airplane's tailskid left the runway while the airplane was gathering speed on takeoff! It had to be taken back to the factory for extensive redesign and reconstruction, which would take until June 30.[19] At 57,338 guilders, the F.VII would be the most expensive Fokker airliner of the 1920s in terms of research and development costs.[20]

After the conflicts over price, this latest mishap further harmed the relationship between Fokker and Plesman, who even went as far as to suggest that Fokker might be mentally deranged.[21] In business, matters would only become worse: on March 19, 1924, following up on the promises made by Fokker ten months earlier, KLM formally approached him and (again) the NVI to come up with a design for a safer, more reliable three-engined airliner. The airline was willing to pay a hundred thousand guilders for such an aircraft. But now, Tony Fokker said he was not

interested: the price KLM had in mind was far too low, and with expected improvements in the engine, he believed a three-engined airplane
would be obsolete upon completion. Hence he had "no intention whatsoever of risking capital for the benefit of KLM."[22]

As it turned out, Fokker was fortunate not to bid for the contract, a decision that would rid him of a potentially dangerous Dutch competitor.
Eager to please KLM, NVI invested all its resources in its three-engined
project, designated the FK-33 after the initials of its designer, Frits Koolhoven. This was a fateful decision. The FK-33 proved costly and disas-
trously unsuitable for KLM, and its failure ruined the NVI.[23] When the
prototype was finally taken over by KLM in December 1925, it needed so
many changes that it took KLM's technical staff until November 8, 1926,
before the FK-33 was ready for operational testing. The aircraft was
taken on a proving flight to Paris, which very nearly ended in a crash. Part
of the wing cover and an engine came loose. In his flight report, the pilot
stated, "I have . . . to emphasize that I have lost all confidence in the construction of this machine, and . . . that it is my opinion that this airplane
is not suitable for any purpose whatsoever. The risk of a severe fracture
after a certain time is too big."[24]

As a result of this fiasco, KLM continued to depend on Fokker. To
make sure no resurgence of domestic competition took place, Fokker and
NVI reached a confidential agreement under which Fokker took over the
financial obligations of the NVI and compensated its shareholders in exchange for a promise from its backers to refrain from any future involvement in aircraft construction.[25] As a result, in 1926 Fokker reemerged as
Holland's only aircraft producer.

His deal with the NVI revealed, once more, an ingrained trait in Tony
Fokker's character. Where business was concerned, he left nothing to
chance and reduced the element of risk to an absolute minimum. In this
case, a neat masterstroke created a monopoly.

Thus, Fokker continued as a producer of aircraft custom-built to
KLM's requirements. In June 1925 a contract was drawn up for five aircraft of the F.VIIA type, an aerodynamically much-improved version of
the original F.VII—the effect of Fokker's taking on a new qualified engineer as chief of research, Bertus Grasé, who had worked for the government's air research bureau, RSL. A "born pilot," Grasé also became the
company's main test pilot until his death from cancer in 1929. Powered
by an air-cooled engine, which offered a better power-to-weight ratio
than the liquid-cooled engine of the original F.VII, the F.VIIA became

Fokker's next civil export success. Fokker built forty-two of them in Amsterdam, and more were constructed abroad, under license.

Once again—by now, almost inevitably—KLM was to be the launching customer for this new type. And yet again, the relationship between Fokker and the national airline was affected by the process of finalizing the sale of five such aircraft. Both parties showed that they had learned little from their earlier skirmishes. As before, the conflict centered on the price Fokker asked for the aircraft, which had not been finally determined in the contract. Emotions and legal threats ran even higher than before, and in August 1925 court proceedings were only narrowly avoided when KLM finally gave in and paid Fokker's asking price.[26]

Tony Fokker was not around to celebrate this victory. He was in New York and far too busy preparing his permanent move to the United States (see chapter 6), while Platz and his team were putting in overtime in Amsterdam to complete the first of the three-engined variants of the F.VII before the end of the month.

That summer the American car manufacturer Henry Ford completed his takeover of the Stout Metal Airplane Company on July 31, 1925—a process that had started late in 1923, when Ford came to the financial rescue of the company's constructor, William Stout. At the same time, Ford was organizing the National Air Reliability Tour—the first of a series of seven, as it turned out—to boost the American public's awareness of aviation's increasing safety, a condition of vital importance to commercial air transport, which was only then coming into existence in the United States. At the same time, the Reliability Tour was intended to provide extra news coverage for the products of the new Stout Metal Airplane Division of the Ford Motor Company. The idea was to hold an aerial *tour,* rather than a race, in which all entrants would complete a circuit of 1,900 miles from Dearborn, Michigan, to Fort Wayne, Chicago, Moline, Des Moines, Omaha, Saint Joseph, Kansas City, Saint Louis, Indianapolis, Columbus, Cleveland, and back to Dearborn.[27] When, early in July, Fokker heard about the tour, he decided that what worked for Ford might also work for Fokker by creating a wide market for his year-old American venture, the Atlantic Aircraft Corporation.

Fokker realized that the three-engined variant of the F.VII discussed with KLM in 1923, but then shelved, could be just the kind of airplane to carry him not only to the forefront of the Reliability Tour, but also of American aviation as a whole. Three engines would reduce the risk of having to drop out of the tour because of engine trouble and give the

F.VIIA (which had just been completed and successfully test-flown at Schiphol by Grasé) extra speed over the predominantly single-engined competitors. Early in July, Fokker sent a telegram to Amsterdam with instructions to equip an F.VIIA with two additional engines forthwith, and he promptly shipped three Wright Whirlwind engines to Holland from the United States.[28] According to Bruno Stephan, it was the last direct instruction in aircraft design that Tony Fokker gave to the Nederlandsche Vliegtuigenfabriek.[29]

On September 3, only eight weeks after Fokker's initial telegram, the airplane was ready. Four days later, Fokker himself, just arrived from the U.S., approved of its flying characteristics and had the aircraft crated and shipped aboard the Holland-America liner *Veendam* to New York. From there, Fokker flew the airplane to Detroit on September 26. Dominating the Reliability Tour, and treating it like the race it was not intended to be, Fokker crossed the finish line first in Dearborn, Michigan, though closely followed by Ford's single-engined Model 2-AT.[30] From then on, the F.VII became Fokker's big success story in civil aircraft construction. Including licensed production in Belgium (SABCA), Britain (Avro), Poland (Plage & Laskiewicz), Italy (Officine Ferroviarie Meridionali), Czechoslovakia (Avia), Hungary (Manfred Weiss), and Fokker's American production line, some 250 planes of the various single- and three-engined F.VII types were built.[31]

The improved version of the three-engined airplane, the F.VIIB, featuring a bigger wing (71.2 feet, against 63.4 feet for the F.VIIA) to provide additional lift, and a 27-percent-higher takeoff weight (11,461 pounds), subsequently led to the development of a third variant of the basic F.VII design. The twin-engined F.VIII of 1927 had a still bigger wing (75.5 feet) and a fuselage stretched 7 feet (to 55.1 feet) for a cabin seating twelve to fifteen passengers. Its design origins, and the pragmatic approach of the Fokker company, showed in its extremely low research and development costs: a mere 6,380 guilders. Bigger still, with an 88.7-foot wing and a 60.8-foot fuselage, and powered by three 450 hp Gnôme-Rhône Jupiter engines, was the F.IX, which came out in 1929 and was developed at a cost of only 29,145 guilders. The F.IX had been intended for KLM's Amsterdam-Batavia service, but the airline felt the F.IX, seating up to twenty passengers, was too big and heavy for the task. Instead KLM ordered a downsized variant, the F.XII, which first flew in 1930. The F.IX was then transferred to KLM's European network. But the F.IX, with operating economics no better than that of the F.VIII—which in practice offered similar seating capacity—remained unsuccessful. Only two were built in Holland, against thirteen F.XII's.[32] These low-cost designs and their limited production runs showed that Fokker's Nederlandsche Vliegtuigenfabriek was slipping toward a crisis. The company kept trying to secure orders for large exports of military aircraft—which suffered from the same lack of research investment and technological progress as Fokker's civil types. Dwindling allocation of funds for research and development was undermining Fokker's market position in aircraft manufacture. Fokker's personal life, along with his combined role as director and the only shareholder of his Nederlandsche Vliegtuigenfabriek, were to threaten the life of the business.

## A SELF-MADE MAN WITH WINGS

If, at the time of his marriage in March 1919 with Tetta von Morgen, Tony Fokker had planned to retire from aircraft construction and spend his days enjoying life and leisure, such plans appear not to have been very deep-rooted. Within a few months, he was engulfed in efforts to launch his Dutch enterprise. He quickly came to be recognized as a leading personality in Dutch aeronautical circles, too, honoring gatherings of the Koninklijke Nederlandsche Vereeninging voor Luchtvaart (KNVvL: Royal Netherlands Aeronautical Society) with his presence. When the first London-Amsterdam flight of Air Transport & Travel and KLM arrived on May 17, 1920, Tony Fokker was the one who pried open the mailbag for the press cameras.[33] A prominent member of the KNVvL's committee for aerial tourism, he spoke out repeatedly on issues such as airport construction and airplane certification, and he became the main sponsor of the committee that organized the first flight from Amsterdam to Batavia in 1924. These and many other commitments began to put a strain on his relationship with Tetta. Fokker, a workaholic, spent an amazing amount of his time on business travel abroad. For a renowned aviator and aeronautical businessman like Tony Fokker, it was remarkable that he preferred to journey in the comfort of first-class trains rather than take a plane.[34] Seldom home for any extended period of time, he was extremely egocentric and had odd habits that made him difficult to live with. Fokker was singularly obsessed with things aeronautical. Conversations outside the subject of aviation did not hold his interest for long, and he often showed a lapse of attention in a disturbingly direct way by closing his eyes and dozing off in the middle of the conversation. Taking a quick nap on the couch was Fokker's favorite way to revitalize.

Even to those close to him, he remained a man of odd contrasts. He refused to be contradicted and hardly ever complimented anyone, even for hard-earned achievements. Paternalistic, pushy, unpredictable, a teaser, Fokker could be arrogant to personnel and servants whom he did not know personally ("They get paid for it, don't they?"), yet he could also be extremely amiable, funny, and generous, if he felt like it.[35] His head of the household and cook, Christine Döppler, who had followed Tetta to Holland as a lady-companion and personal attendant, and who remained in Fokker's employ until 1926, had a lifelong infatuation for him.[36] To his closest friends, such as the Dutchman Wim van Neijenhoff, he showed a different, personal side. They knew him as an unpretentious, homely man who dressed carelessly and preferred to walk around in suits that were too big for him, stuffing his pockets with a variety of odd objects he happened to pick up. He loved animals, especially dogs, and had a number of canine companions in various phases of his life. Apart from a distinct preference for rising early in the morning to put in some hours of work before breakfast, he kept no regular time schedule and was never on time for appointments, if he remembered them at all. Arriving home after midnight, sometimes even bringing guests for dinner at that hour, was not especially unusual for him either. And though he enjoyed wining and dining his guests, he himself was a teetotaling nonsmoker, addicted to sweets. Ever since his brief enlistment in the army, his mother sent him parcels of candy and cookies wherever he went. He was especially partial to Haagsche Hopjes, a typically Dutch candy made from flavored molasses and butter, and he was forever munching them.[37]

No family man, Tony Fokker preferred to spend the free time that he had tinkering with the engines of his flashy Lancia sports cars or boating in the company of friends such as Prince Hendrik, with whom he went seal hunting in the shallow tidal waters of the Waddenzee, north of Holland.[38] Trips in his two-masted sailing yacht, *Hana,* tended to be all-male affairs. Fokker was an expert yachtsman and owned several motorized and sailing yachts at any one time. His favorite boat, the motor yacht *Honey Moon,* was built to his own specifications. Boating trips on Holland's many watercourses offered rare occasions of true diversion and some degree of marital privacy as well. In the early years of their marriage, Tony and Tetta savored such moments of being away from the pressures of life. On one excursion, they had fun practicing waterskiing on an improvised, self-made board. Sailing, preferably with the salty sea wind blowing in his face, was Tony's favorite relaxation throughout his life. His other big passions were photography and filmmaking. At special events, Fokker seldom turned up without his camera.

But there was precious little time for marital privacy, and Fokker was not the type for it, either. His blatantly flirtatious behavior around women, whether in his employ or by chance encounter, put a further strain on the marriage.[39] Over the years, he and Tetta gradually became estranged. In the summer of 1922, she stopped accompanying him on his frequent foreign trips in Europe and the United States. On October 11, 1923, the marriage ended in divorce. Tetta went back to Germany, disillusioned. She would not marry again for many years, finding little happiness in her solitary existence and gradually declining financial means.[40] The only woman Tony Fokker felt close to throughout his life was his mother Anna. Though he seldom wrote her—writing was never Fokker's favorite occupation—he did telephone her every once in a while, which, from the United States, was both unusual and expensive in those days. On special occasions he made sure that she received flowers and presents.[41] His brother-in-law, Geert Nijland, thought she was the only one who really understood him.[42]

After his divorce, Fokker began to spend increasingly less time in Holland. During the winter months he could be found in the exclusive Swiss resort of Saint Moritz, skiing and sliding his skeleton sled (a now-extinct cross between a child's sled and a bobsled) in prone position down the mountain slopes at breakneck speeds. After several years at the fashionable Palace Hotel in Saint Moritz, he bought his own chalet there from the Cremer family. Oberalpina, as the place was known, was a large, permanently staffed house on a mountain slope outside of town, decorated with beautiful, though somewhat dark, rustic wood paneling in nineteenth-century style. Each winter Tony Fokker took personal delight in driving his custom-built Ford snowplow to clear the chalet's access road after a snowfall.

At Oberalpina, Fokker liked to entertain family, friends, and business acquaintances, sometimes as many as fifteen guests at one time. He had special porcelain plates made to hand out as keepsakes upon their departure. Otherwise, he retained many characteristics of a mischievous adolescent, who delighted in shocking new arrivals by welcoming them while lying in his bathtub, covered with soapsuds.[43] On another occasion he put a low-voltage electrical charge on the carpeting in his room, normally out of bounds for family members and visitors, invited some of the female guests to come and look at his private quarters, and took great pleasure from seeing them wince and giggle as they touched each other, thus closing the electrical circuit.[44]

Despite such demonstrations of playfulness, Fokker's youth was slowly slipping away from him. The years of flying in open cockpits were beginning to make themselves felt. Increasing sinus trouble and the severe headaches accompanying that condition led him to travel to Berlin for an operation to clean out the sinus cavity. In another example of the boyish mischief so typical of him, Fokker had the grayish substance the surgeons removed from his head put into a tiny glass jar, which he liked to flash at disgusted onlookers and tell them where it came from.[45] Nevertheless, the operation brought only temporary relief. As the years went by, he developed chronic cold symptoms and never went outdoors without his Kleenex tissues.[46]

Little by little, the tinkering constructor and test pilot found himself "flying a businessman's desk." His American endeavor commanded ever more of his attention and time. When his father, Herman Fokker, died on December 19, 1924, he was tied up in business in New York and unable to attend the funeral. In September 1925 he moved to the United States permanently, firmly believing that the future of aeronautics lay in America's vast distances and proverbially unlimited opportunities. In 1926 he bought an roomy apartment on the fifteenth floor of an externally nondescript but internally luxurious apartment building on the corner of Manhattan's Riverside Drive and West 101st Street, overlooking the Hudson River. When, in January 1928, he arrived in Holland to receive a series of public awards for his achievements in aeronautics and business, a year and a half had already passed since he had last visited the homeland. When his Dutch company celebrated its tenth anniversary in July 1929, he was not even present. The United States mesmerized him; it was where he wished to focus his attention and his resources. His Dutch enterprise suffered as a result and gradually slid into decline.

## IN THE EYE OF THE BEHOLDER

In the 1920s, the number of Fokker's European civil aircraft sales was relatively modest, a reflection of the size of the market, but having as much to do with air transport itself in those days. The on-time performance of airlines was low. For a long time air transport was characterized by the adage, "If you have time to spare, go by air!" Flying was at once expensive and risky, and air transport developed slowly as a result. Delays because of engine and general mechanical trouble, and because of adverse weather conditions, were frequent. The problem was especially bad in winter. As late as the winter of 1923–24, one-third of KLM's scheduled flights were canceled because of weather or breakdowns, compared to only 2 percent during the previous summer season.[47]

In single-engined aircraft such as the early commercial Fokker models, breakdowns could be quite hazardous. On October 19, 1923, KLM pilot Iwan Smirnoff was flying his Fokker F.III from Amsterdam to London when a defunct waterpump made the engine stop dead over the English Channel. Preparing to ditch the plane, Smirnoff was extremely lucky to find himself over the Goodwin Sands sandbank at low tide and managed to put the crippled plane down on it. He and his three passengers then spent a scary two and a half rain-soaked hours on the wing of the plane, watching the tide rise, as their emergency flares failed to attract attention from the crew of the nearly East Goodwin lightship. The four distressed men saw ten ships sail by before the British freighter *Primo,* bound for the French port of Rouen, heeded their distress signals. A lifeboat was lowered but failed to reach the sandbank because of heavy winds. Smirnoff and his passengers had to swim some sixty feet in the cold North Sea water to climb on board and then help to row the sloop back to the *Primo.* It took about an hour in high seas, which eventually smashed the lifeboat against the hull of the freighter. All were saved, but it was 5:30 P.M. before the *Primo* could continue its journey—3½ hours after the plane had touched down on the sandbank. By then, the aircraft had disappeared beneath the waves.[48]

Despite its peculiar development process, the F.III turned out to be one of the more successful early passenger airplanes, and even with its single engine, fairly reliable. It was sold to a number of European, particularly German, airlines. Deutsche Luft Hansa operated eight of them; Deruluft used the type on its long-distance services between Berlin, Königsberg, and Moscow, extending its service to Leningrad in 1928. The F.III also raised the standard of passenger comfort over that of its predecessor, the F.II. Nevertheless, there was still ample opportunity for improvement. One early passenger, aviation journalist Henri Hegener, later recollected:

> Like all the early passenger planes, the F.III had . . . shortcomings too. In a brochure it was said the air circulation and heating was taken care of and in the winter warm air could be conducted into the cabin from the

engine. But whether it was summer or winter, what the engine produced as "warm air" was in effect an evil smell! Airsickness, chiefly caused by the bumping as a result of low flying, often occurred. The airline companies showed understanding for the suffering of their clients by supplying each traveller with an aluminium container which could be closed with a lid. Later, the well-known paper bags which one could deposit out of the window were introduced.[49]

In the cabin, the noise level was tremendous. Early passengers and pilots alike emerged from their flights deafened and dazed, no matter how much effort was spent on soundproofing. One early Dutch passenger, who remarked enthusiastically, "Not everyone will see this as a disadvantage; one can also *be happy in silence,*" was the exception.[50]

Over the years numerous attempts were made to raise the standard of cabin comfort, but even Fokker's last numerically successful airliner, the trimotored F-10A, which was manufactured in his American plants from late 1928, had a cabin that was considered cramped and low. A person of normal height could not stand upright in it, and the noise was so bad that conversation was "almost impossible."[51]

Fokker's European market position was strengthened by the first successful flight from Amsterdam to Batavia, between October 1 and November 24, 1924. A mixed crew made up of KLM's chief pilot, Abraham Thomassen à Thuessinck van der Hoop; Jan van den Broeke (mechanic); and Air Corps lieutenant Hans van Weerden Poelman (copilot) flew the rebuilt prototype of the Fokker F.VII all the way to the capital of the Dutch East Indies, a trip the single-engined plane stood up to surprisingly well after an early engine breakdown led to a forced landing near Plovdiv in Bulgaria.[52] The flight had been organized by the specially formed Comité Vliegtocht Nederland-Indië (Committee for the Netherlands-Indies Flight), made up of high-ranking Dutch industrialists, all believers in the gospel of aviation, and sponsored by 113 Dutch companies and private individuals. Besides being the primary sponsor (with 15,000 guilders),[53] Tony Fokker played a major role in the technical preparations for the flight. He realized that the event might well provide him with free publicity and help his civil sales if it went well. Fokker found that record flights definitely paid off, and in the 1920s he would sponsor and provide planes for a number of such undertakings.

The first such flight came about in the aftermath of Fokker's participation in the Ford Reliability Tour. The safety aspects of the three-engined airliner not only attracted new airline customers for Fokker's commercial models, they also proved to be vital in arousing interest in his airplanes for special flights. Some months after the Reliability Tour, on February 13, 1926, the United States naval commander and polar explorer, Richard E. Byrd, approached Fokker with a bid to buy his F.VIIA-3m display airplane and use it on a planned flight over the North Pole. At first Fokker was less than keen to sell the aircraft. He was well aware of its value as a demonstration model on the emerging airline market in the United States. Besides, the United States Army had also shown interest in the aircraft, so he was loath to risk losing it in an arctic adventure. Yet he also sensed that if Byrd's expedition were successful, backed as it was by influential government institutions and private individuals such as John D. Rockefeller and Edsel Ford, it would generate much free publicity for his products. After further deliberations, and no doubt having checked into the background of the Byrd expedition, Fokker agreed to sell the aircraft for $40,000 on the condition that the name Fokker be left clearly visible on the aircraft.

This was a hefty price, but Fokker had a good bargaining position, having already sold and delivered another trimotor for a similar venture. The Australian Army pilot and polar explorer, Capt. George Hubert Wilkins, was organizing the Detroit Arctic Expedition, backed by Detroit businessmen and private individuals, some 80,000 in all (including a surprisingly high number of schoolchildren). Late in 1925 Fokker had met Wilkins in Detroit when the latter was trying to raise funds for his proposed arctic flight from Point Barrow, northern Alaska, to Spitsbergen, the group of islands halfway between Norway and the North Pole. Wilkins's proposed adventure appealed to Fokker, who suggested that he fly the latest version of the trimotored F.VII, then under construction in Amsterdam, with a wingspan increased from the original 63' 4" to 71' 2". When Wilkins made some cautious remarks about funding for the expedition, Fokker told him not to worry: if the Detroit money did not come through, he would personally foot the bill for the three-engined plane. Wilkins was delighted.[54] He now had two aircraft, the other a single-engined F.VIIA that he had bought from an American owner. The single-engined airplane was later christened *Alaskan;* the three-engined plane, *Detroiter.* Both were fitted out at Fokker's expense at his Hasbrouck Heights assembly plant, just south of New York. Late in January 1926, preparing for what was to be the first race to overfly the North Pole, Wilkins shipped his two aircraft from Seattle to Seward, Alaska, for reassembly. But his pilots, unfamiliar with Fokker aircraft, severely

damaged both machines on their respective first test flights on two consecutive days. Repairs would obviously take time.[55]

Byrd had originally pondered the idea of using an airship for his endeavor, like the Norwegian explorer Roald Amundsen, another competitor in the race, and was in a hurry to acquire a reliable airplane that
could do the job with the least time for preparation. He named his aircraft *Josephine Ford* after Edsel Ford's three-year-old daughter. Taking off
from the island of Spitsbergen, Byrd and his pilot, Floyd Bennett, returned as "the first to overfly the Pole," claiming to have dropped the
American flag there on May 9, 1926, despite failure in one of their three
engines. Byrd and his crew returned to the United States to be greeted by
a traditional New York ticker-tape parade in June.[56]

The publicity generated by Byrd's flight boosted Fokker's American
sales, and when Byrd approached Fokker late in 1926 to buy an improved trimotor (originally designed as a United States Army transport
and thus known by its military designation, the C-2), he had rather less
trouble reaching an agreement with Fokker. Byrd wanted to fly the new
airplane in the big, much-publicized aerial contest of 1927, the first nonstop transatlantic crossing to Paris. The winner would take home the
Raymond Orteig Prize, $25,000, which had been announced in 1920.
Several competitors were preparing for the big hop, giving the project a
sense of urgency. Fokker had the wing for Byrd's plane built in Holland
to save time. Like Wilkins's aircraft, it had an increased wingspan. On
April 20, 1927, the machine, named the *America,* was rolled out for its
first flight. Byrd, Bennett, and their radio operator, George Noville, demanded to be taken up, and Fokker, acting as test pilot, grudgingly consented. This was a regrettable decision, because with the new wing, and
little fuel on board, the aircraft proved nose-heavy, causing it to overturn
on landing as the propeller of the central engine hit the ground. Bennett,
in the copilot's seat, badly fractured his leg and suffered multiple cuts
from broken glass. Byrd broke his arm. Noville suffered internal injuries.
Only Fokker, with his eternal test-pilot's luck, clambered out of the
wreckage practically unhurt.[57] Though the airplane could be repaired,
the incident caused personal friction between Byrd and Fokker, which got
worse as Byrd, worried about the financial consequences of a further disaster,[58] took much longer to prepare for the transatlantic flight than
Fokker thought necessary. This led to a nasty row between the two men
in the *America*'s hangar at Roosevelt Field one morning. Fokker wrote,
"Three times we pushed the completely equipped *America* to the top of

the fifteen-foot hill . . . which I had devised to give the *America* the equivalent of 500 feet more run. I sent word each time to the Garden City Hotel that all was in readiness, but the flyers failed to show up. . . . Fed up, I went aboard my yacht in the Sound."[59]

On June 29, more than a month after Charles Lindbergh's epic transatlantic flight, Byrd; his pilots, Bernt Balchen and Bert Acosta; and his radio operator, Noville, finally made their hair-raising, heavily overloaded takeoff, failed to find Paris in the fog after their compass and radio broke down, and crash-landed the *America* two hundred yards off of the French coast of Normandy, in front of the lighthouse of Ver-sur-Mer, at 3:30 A.M. on June 30. Byrd managed to salvage the mail from the wreckage, only to discover later that sea water had soaked the letters and loosened the stamps. The animosity between Byrd and Fokker created by the events around Byrd's "scientific" flight was intense. In Fokker's eyes, Byrd's indecisiveness had cost him his, Fokker's, rightful place in the annals of aviation history as the constructor of the first airplane to fly nonstop across the Atlantic. The episode lingered in Fokker's memory. In *Flying Dutchman,* he made a point of expressing his low esteem for Commander Richard Byrd.[60]

Fokker's involvement in other special flights were less confrontational. Back in Holland, plans for long-distance air services—one of the incentives behind the founding of KLM—were just emerging from their preliminary stages. In aeronautical terms, the world was certainly shrinking in 1927, and Tony Fokker's planes had a lot to do with that. In the last week of February, the American millionaire Van Lear Black, owner of the *Baltimore Sun* newspaper and several banks and mining companies, walked into KLM's London office with a request to charter one of its Fokker aircraft to fly to Cairo, Egypt, to surprise a friend who was there on a sailing trip. KLM, which at that time was fighting off imminent bankruptcy while negotiating a new subsidy agreement with the Dutch government, relished the chance for unexpected extra earnings and gladly cooperated. Between March 1 and June 14, Black toured Europe in one of KLM's Fokker F.VIIA planes. After Lindbergh's success, an enthused Black proposed that KLM fly him back to Baltimore across the ocean. This quickly proved to be technically unfeasible, but Plesman *was* able to offer him a replacement adventure: a trip to the Dutch East Indies.

In the exceptionally short time of two weeks, Black's plane was overhauled and readied for the flight. On June 15, KLM's pilots Gerrit Geyssendorffer and Johan Scholte, and their mechanic, Karel Weber, took

off with Black and his butler, Harry Bayline. They arrived at Batavia's Ke-
majoran airfield on June 30, where the airplane was turned around and
flown back to Holland. After a remarkably uneventful five and a half
weeks, they touched down safely in Amsterdam on July 23—an unexpected success.[61] Though Fokker had nothing to do with this flight, its
success spurred him to agree to a proposal from the Netherlands Air
Corps pilot George Koppen for a record-breaking return flight to Batavia,
carrying mail. Fokker made available a brand-new three-engined F.VII
(dubbed *Postduif:* carrier pigeon) for the endeavor, which was further
sponsored by the Comité Vliegtocht Nederland-Indië. Twenty-two days
after takeoff on October 1, 1927, the *Postduif* and its crew (Koppen, G.
Frijns, and S. Elleman) landed safely back in Amsterdam—proof of what
the new three-engined Fokkers could do.[62]

Then, in the spring of 1928, the Australian airman Charles Kingsford-
Smith approached Fokker in New York to enlist support for his intended
round-the-world flight. For this endeavor Kingsford-Smith had bought
the cracked-up *Detroiter,* which Wilkins had shipped back from Alaska
to Seattle,[63] repaired it, and took off with his fellow crew members
Charles Ulm, Harry Lyon (navigator), and James Warner (radio operator)
for Hawaii from Oakland, California, on May 31, 1928. From there,
they flew their aircraft, now named *Southern Cross,* in stages to Sydney,
Australia, where they arrived ten days later on June 10.[64] To the delight
of the aeronautical world, the flamboyant "Smithy" did not rest there but
later continued his westward flight and arrived in Amsterdam in July
1929.

In Amsterdam the *Southern Cross* was taken apart on Fokker's instructions for a free major overhaul, and not a moment too soon. Even
casual inspection of the aircraft showed that the dihedral angle of the
wing had visibly increased. To the dismay of Fokker's chief engineer,
Marius Beeling, it was discovered that the (wooden) forward wing spar
had developed serious cracks, which had been repaired simply by nailing
some planks to it.[65] After the *Southern Cross* was rebuilt, Kingsford-
Smith, now joined by one of KLM's senior pilots, Evert van Dijk;
J. Patrick Saul (navigator); and John Stannage (radio operator) made the
westbound Atlantic crossing between June 24 and 26, 1930. In New
York, the pilots were generously received by Fokker, who organized a fes-
tive get-together for the long-distance aerial explorers at his house in
Alpine-on-Hudson. After that, Kingsford-Smith flew on to Oakland, his

original starting point. The *Southern Cross* was shipped to Australia to be exhibited eventually in the aviation museum at Canberra.

These well-publicized flights, and the one that had taken the American "Lady-Lindy," Amelia Earhart, and her pilot, Wilmer Stultz, in their Fokker *Friendship* across the Atlantic to Wales in June 1928, did more than just prove the reliability of these planes and their possible use for scheduled air transport. They established Tony Fokker's name worldwide as a producer of highly successful commercial airliners. But despite his usual sharp business instincts, Fokker closed his eyes to the long-term potential that this universal acclaim created and the demand for new, path-breaking designs that would inevitably follow. Instead, the development of his commercial models suffered from the same basic problem as his military designs: the intellectual limitations of their designers, and Fokker's reluctance to invest substantial sums of money in fundamental research and development, both essential to taking aircraft development beyond his own hands-on understanding.

# 6

## THE AMERICAN ENCHANTMENT

After the war, substantial numbers of Germany's top-rate fighter aircraft, predominantly Fokker D.VII's, were shipped across the Atlantic for operational and industrial evaluation. It has been estimated that as many as 142 of them were brought into the United States as war booty; another 20 went to Canada.[1] Although the D.VII's were not put into service with active squadrons, their evaluations did generate American interest in other types by the same, now ex-German, manufacturer. Through the military and naval attachés of the American Legation in The Hague, the U.S. Army Air Corps bought one D.VIII fighter in 1920, and one two-seat Fokker C.I observation and light-bomber aircraft from Fokker's war surplus stock. The U.S. Navy bought three C.I's.[2]

Such interest was most gratifying for Tony Fokker, who felt ill at ease in his newfound homeland. When he was approached in September 1920 to come to the United States and see what was being done with his airplanes there, he did not take long to accept the invitation, which had been in the pipeline since late 1919.[3] Tony, Tetta, and Tetta's brother, Hans Georg von Morgen, arrived in New York aboard the liner *Noordam* of the Holland-America Line on November 10, 1920. Fokker was in high spirits. In an interview with the *New York Times,* he admitted having come to the States to explore new markets for his aircraft, stating that he expected regular transatlantic services, flown with large, all-metal landplanes, to emerge in the next five to ten years.[4] To prepare the way for him in the U.S., Fokker was also further accompanied by a fellow Dutchman, Robert (Bob) Noorduyn. Three years Fokker's junior to the day, Noorduyn already had a lengthy career in aircraft construction behind him. Starting as a draftsman for the British firm of Sopwith Aviation in 1913, he had then joined up with the Dutch designer, Frits Koolhoven, at Armstrong Whithworth (1914–17) and subsequently at British Aerial Transport (1917–20) as chief draftsman and supervisor of construction. Noorduyn, who spoke fluent English as a result of his seven-year sojourn in Britain, would be a real asset to Fokker in developing his American interests and was to occupy a key position as Fokker's general manager there until 1929.

Confirmation of U.S. interest in Fokker aircraft did not take long. A month after his arrival, he signed a contract for delivery of two big transport aircraft, of which he must have made brief sketches as a result of conversations with U.S. Army officials while visiting the U.S. In true Fokker style, the idea was simple: a greatly enlarged version of the F.III design that had just then been completed in Holland—the F.IV. Hoping that interest in the F.IV might be just the tip of the iceberg, Fokker set up the Netherlands Aircraft Manufacturing Company as his sales agency in New York before returning to Holland on January 8, 1921. Fokker's foothold in the New World was rather informal at first. Office space was rented on New York's fashionable Fifth Avenue, where Bob Noorduyn shared responsibilities as Fokker's American representative with Fokker's long-time friend and partner Frits Cremer, whose father now occupied the position of Netherlands ambassador in Washington. But after ten years of collaboration, the Fokker-Cremer association came to an end. Cremer, more a pilot and bon vivant than a salesman, left in 1921 and went to live in Milwaukee.[5]

Fokker was easily captivated by the seemingly boundless opportunities the U.S. was famous for. America was what Holland was not: following a brief depression at the end of 1920, the intertwined developments of a booming economy and consumerism were rapidly transforming the physical landscape of cities and village communities alike into ever-taller buildings and paved roads. Spending on advertising more than doubled between 1918 and 1929. Much of that growth came from the mass production of consumer goods, such as cars, washing machines, refrigerators, and radios sold to a broad section of society. Manufacturing output of durable consumer goods almost doubled between 1921 and 1929.

Though inequality in prosperity increased, credit brought luxury goods within the reach of the average worker.[6] The private car was emerging as the symbol of an economic and technological revolution. With it, the new phenomenon of mass culture appeared. On November 2, 1920, an amateur broadcaster in Pittsburgh, Pennsylvania, identified by the call sign KDKA, started the nation's first radio broadcasting station, playing phonograph music and requests. Obliterating distance, it proved an extremely popular medium for entertainment and information alike: more than three million radios were sold in the first two years alone.[7] Fokker also bought one, and radio listening became one of his favored forms of relaxation.[8]

During the two months of his trip to the U.S., Fokker traveled around a lot, speaking on aviation matters to select and enthusiastic audiences. In a letter to his house staff in Amsterdam, he commented on the luxuries of American hotels. Central heating was so efficient, he said, that he was able to travel with nothing more than shirts and ties, a smoking jacket, and some other loose garments, despite the wintery cold.[9] He returned to Holland (and his quarrel with Plesman over the F.III order) in high spirits.

Further orders from the American military trickled in during 1921, justifying Fokker's decision to have Noorduyn organize a permanent New York sales office. Two two-seater D.VII conversions were shipped to the U.S. in 1921 for army testing, along with the single D.VIII. The company also managed to interest the U.S. Army in an improved version of the D.VII, the D.IX (known by the army designation of PW-6: pursuit, water-cooled, model 6), but because the airplane was of foreign design and could offer only marginally better performance than the regular D.VII, only the prototype was sold. It was shipped to the United States in June 1922. A contract was also signed for a further development of the D.VIII, curiously designated F.VI by Fokker (PW-5 by the Army Air Service). Two prototypes were delivered early in 1922, followed by ten production models in May of that year. Three D.XI's (PW-7) were contracted in August. By then the two big F.IV's (army designation T-2/A-2) had also been shipped to New York.[10]

Not all of Fokker's early endeavors in the U.S. were successful, however. In July 1921 the first of two F.III airliners was exported to the U.S., but despite a demonstration tour with the well-known pilot Bert Acosta at the controls, no buyers were found.[11] Unlike the situation in Europe, where the development of civil aviation was largely characterized by government-subsidized national airlines that carried passengers, mail, and cargo, nonmilitary flying in the U.S. was restricted to airmail services operated by the U.S. Post Office and shoestring operations of "barnstorming" flyers that sought to attract the public's attention by performing a variety of dangerous stunts in ramshackle, war-surplus aircraft. This being the case, there was simply no market for the airliners the Nederlandsche Vliegtuigenfabriek was trying to sell. With no big military orders materializing as a result of congressional "buy American" pressure, Fokker's venture seemed to be heading for an impasse.

Yet early in 1922 things suddenly started to look up. In December 1921 the American war ace and relentless organizer of a modernized U.S. air force, Gen. William (Billy) Mitchell, and his aide, Capt. Clayton Bissell, boarded the transatlantic liner *Rotterdam* for Europe, where they were later joined by the self-taught aeronautical engineer Alfred V. Verville. Their mission, "smacking as much of adventure as of hard work," as one author put it, was to "obtain complete and exhaustive information" on every possible aspect of European aviation.[12] After several weeks of combining their fact-finding tour with the pleasures of Parisian society events, Mitchell's party headed south in January 1922, visiting aeronautical establishments in Italy and Germany before stopping at the Nederlandsche Vliegtuigenfabriek late in February.

The visit had far-reaching consequences. Fokker pulled out all the stops to make a positive impression on his guests, who might prove instrumental in furthering his American exports. He pampered them, drove them around in his car, arranged for them to be received by Queen Wilhelmina. The time was well spent. Mitchell was positively impressed by the various aircraft shown to him. He bought the prototype C.IV single-engined light two-seater reconnaissance bomber that Fokker demonstrated, eventually leading to eight such planes being shipped to the States (as CO-4's). Mitchell's say was also vital in securing an order for ten PW-5 parasol-winged monoplane fighters. He also ordered a training aircraft of similar layout that Fokker showed him, the S.I (army TW-4). Something of a hands-on technician himself, Mitchell was given a direct demonstration of Fokker's empirical working practices. Bissell was given the opportunity to test the new T.II torpedo bomber, a three-seat single-engined floatplane with a low cantilever wing that Fokker was building for the U.S. Navy, and commented that the airplane's tail controls were sluggish. The following morning, Fokker had the remedy worked out: mechanics cut off a three-foot section of the tail and welded the metal

frame back together. This solved the problem.[13] But even more conse-
quential in the long run, Mitchell urged Fokker to carry out his intentions
of coming to the United States and start his own factory there. To his superiors, Mitchell stated, "This man has capabilities which are unlimited
in the development of aircraft."[14]

Was there a tacit understanding between Billy Mitchell and Tony
Fokker on further U.S. military contracts if Fokker started up an American manufacturing plant? Certainly, Fokker had endeared Mitchell to
him. They shared a single-minded unorthodoxy—a hatred of aeronautical bureaucracy—and were committed to furthering aviation. On May
13, 1922, three months after Mitchell's visit to Amsterdam, Fokker left
for New York aboard the liner *Nieuw Amsterdam,* in which he and Tetta
occupied the forward state room. It would be their last trip together.
Right up to the moment of departure, Fokker had been tied up with business. They only arrived on board at 2:00 A.M., when the ship was await-
ing high tide before passing through the sluice gates of the North Sea
Canal at IJmuiden, six hours after the gangplank had been hauled up for
the night in Amsterdam. But as Fokker jokingly wrote to his mother, "A
good name is a good key!"[15]

On its aft deck, the ship carried the first two Fokker T.II torpedo floatplanes, which the navy had ordered for trials at Anacostia Naval Air
Station in Washington, D.C. Fokker would be present at the trials, where
his aircraft competed against American designs by Curtiss, Stout, and
Douglas, and a British entry by Blackburn. But the trip had more to do
with Mitchell's urging to set up an American factory than with landing
the navy production contract. To Tetta's dismay, Fokker would spend
practically the full two months of their American sojourn talking to various business interests about raising funds for such a plant. He also spoke
out repeatedly against the disorganized state of civil aviation in the
United States, which, already handicapped by a shortage of suitable airports across the country, also lacked legal regulation, enabling competitive practices that endangered safety.[16] The worldwide fame of the name
Fokker enabled him to carry his arguments to the highest levels. On July
6, 1922, the day of the navy trials of the T.II at Anacostia, Fokker was received by President Warren Harding at the White House, where he managed to show up an hour and a half late. Nevertheless, the Netherlands
chargé d'affaires, J. B. Hubrecht, who accompanied Fokker, reported, "It
was remarkable to witness how this young countryman managed to express his views about what needs to be done in the aeronautical legisla-

tive field in this country to the Head of State, with conviction, and yet not without modesty."[17]

Fokker's arguments were not new ones. They merely reflected an industry-wide concern that federal legislation was a vital necessity for further development of civil aviation in the United States, where, contrary to the situation in the European countries, no legal regime existed. Pilots needed no license; goods and people could be carried regardless of suitability of equipment or safety standards. Civil aircraft did not even need certification of airworthiness because there was no government agency assigned to such a task. Even liability in cases of aircraft accidents remained uncharted territory in law. Moreover, with war-surplus aircraft for sale for as little as $300, and plenty of pilots traveling around the country, willing to risk their lives day after day in hair-raising barnstorming demonstrations, aviation in the U.S. was primarily associated in the public's mind with danger, not transport. To combat this perception, the aviation industry demanded legal action. It was well understood that any possible development of scheduled commercial air transport, of considerable importance as a potential new market for the withering American aircraft industry, would depend on certification, inspections, and a good safety record.[18]

Tony Fokker, therefore, had a strong reason to pursue the issue of legislation, which would have a direct bearing on the commercial opportunities of the American company he was trying to form. When he left again for Holland on July 22, confident of being well on track toward achieving this objective, he told the *New York Times* journalist who saw him off that he expected to return to the United States soon and start building airplanes in North America.[19]

## ATLANTIC BREEZES

The year 1923 was important for the Fokker company. With exports to the Soviet Union at their peak (338 aircraft), orders for the secret German air force in the pipeline, and the design and construction of Fokker's important F.VII civil model in progress, business was booming in Holland.

At the same time, his German company, Schweriner Industrie Werke, was rapidly winding down to final liquidation. So was his other wartime legacy: his marriage with Tetta von Morgen. On July 30, divorce proceedings before the district court of Amsterdam were finalized. Tony

Fokker was once again a "free man," about to embark on a completely new course in the United States. There, he intended to cash in on the success of the first nonstop coast-to-coast flight, by the Army Air Corps lieutenants John McCready and Oakley Kelly, in one of the two F.IV transports (May 2–3, 1923).

On December 18, 1923, Fokker arrived in New York from Southampton as a stateroom passenger on board the luxurious Cunard liner *Berengaria*, having spent a few weeks in Britain discussing trends in aeronautical construction. Four days earlier, on December 14, the articles of incorporation of his new American venture, the Atlantic Aircraft Corporation, had been officially approved by the state of New Jersey. As in his Dutch enterprise, Fokker had prudently left the official procedures to his business associates, in this case, Frank Ford and George Davis (of the consulting and engineering firm of Ford, Bacon & Davis); and Maj. Lorrilard Spencer, a New York businessman, former president of the Wittemann Aircraft Company (1916–17), and a decorated World War I combat veteran. Capitalization stood at around $800,000, with Fokker as the largest shareholder. To shield the new company from the congressional "buy American" requirements then in effect, Fokker initially remained in the background as "chief engineer." Spencer held the title of president, and Noorduyn held the somewhat undefined position of company secretary and general manager.

A British engineer, A. Francis Arcier, was appointed "chief engineer and factory manager," though his letter of appointment made it clear that Tony Fokker called the shots at Atlantic.[20] When Atlantic actually started business in May 1924, it took over the assets of the then defunct Wittemann-Lewis Aircraft Corporation and leased the factory buildings it had constructed only four years earlier on the 800-acre airfield at Teterboro, near Hasbrouck Heights, just across the Hudson River from New York City. For its time, it was a modern factory—small, but equipped for woodworking and metalworking. It had its own modest wind tunnel and a meteorological and radio station. Arcier, the manager and consulting engineer at Wittemann, had witnessed its collapse under the financial burden of a U.S. Army Air Corps contract for a big six-engined bomber prototype, named Barling after its principal designer, after Lewis had bailed out in 1922.[21]

Wittemann also had a contract to rebuild a number of old de Havilland DH-4 two-seater biplanes for the U.S. Post Office, part of a large, federally funded program of modernizing these standard war-surplus planes for continued low-cost use by the army and the post office. Atlantic was able to open for business because it was granted a large slice of this trade—Mitchell's reward for convincing Fokker to start up an American factory.[22] Atlantic received an order to modernize 100 (later increased to 135) DH-4 fuselages, replacing the original wooden framework with Fokker's more durable welded steel tubing construction.

Atlantic started out slowly. Six months after the company opened, it still employed only fifty-three people, the majority of them taken over from Wittemann. Fokker personally supervised the initial stages of the DH-4 rebuilding program, spending much time on the factory floor— the kind of direct involvement he liked best—giving instructions about structural changes, cockpit layout, and so forth. With such a small work force, the reconstruction program went slowly, and the DH-4 order remained uncompleted until 1926.[23]

There was a good reason not to hurry. As U.S. Army imports from the Nederlandsche Vliegtuigenfabriek threatened to dry up, Fokker needed Atlantic as a sluice through which he could continue to export his products from Amsterdam. It was an odd arrangement, because the financial structures of the Dutch and American companies were kept entirely separate. Fokker himself formed the only link between the Nederlandsche Vliegtuigenfabriek in Amsterdam, the Netherlands Aircraft Manufacturing Company in New York, and the Atlantic Aircraft Corporation in Hasbrouck Heights. Though he needed to keep the Dutch and American finances apart, Fokker never quite came to grips with this arrangement and readily instructed his Dutch company to provide the services, parts, and planes he needed at Atlantic. He insisted that this kind of unremunerated deal was just like transferring money from his left pocket to his right, when he was actually doing himself—practically the sole owner in Holland—a disservice to the ultimate benefit of his American shareholders.[24]

Despite Fokker's claims that Atlantic would soon be turning out up to three hundred aircraft a year,[25] the company's development was frustratingly slow, and, had it not been for Mitchell's DH-4 contract, it might have gone out of business. In 1924 a bid for an army training aircraft (the S-3 biplane), loosely based on two similar types being built in Holland (the S.III and S.IV), came to nothing, just as a mail-plane development of the C.IV for the U.S. Post Office failed to win a production order against the far cheaper DH-4 modifications.[26] With the almost complete absence of scheduled airline operations in the United States, civil sales were also not forthcoming.

Prospects started to look up in 1925, when President Calvin Coolidge signed the Air Mail Act into law on February 2. This essential step toward federal involvement in stimulating civil aviation through privatization of airmail carriage had been a long time coming.[27] The Air Mail Act authorized the U.S. Post Office to contract domestic commercial airlines to carry mail—at rates that effectively amounted to a subsidy. Introduced in 1924 by the Republican congressman for Pennsylvania and chairman of the House Post Office Committee, Clyde Kelly, it was specifically intended to further the development of commercial aviation in the U.S.[28] Fifteen months later, it was followed by the Air Commerce Act (May 1926), which formalized the legal structure within which civil aviation was to develop.

Unable to determine where the future of his interests really lay, and with prospects of further military orders declining as Billy Mitchell's star was losing its glitter, Fokker shuttled restlessly between the U.S. and Europe on his favorite transatlantic liners, the *Berengaria* and the *Mauretania*. Fokker loved these journeys, exulting in the care and copious luxury the Cunard Line bestowed on its stateroom passengers. Fokker was partial to good food and had a keen eye for the splendid blend of technology and aesthetics that these ships offered. Such sea voyages beguiled those who could afford them and offered temporary deliverance from the rigors of everyday life by keeping people beyond the reach of business. Fokker's habit of ocean travel was probably motivated by a desire to escape. Besides, frequent journeys between the Old World and the New fit his carefully groomed image of cosmopolitan eccentricity. Nevertheless, during the early part of 1925, he only found real peace in his Saint Moritz chalet, playing in the snow with specially invited friends and family, for he was not a man who stood up to solitude particularly well.

When Henry Ford announced his Reliability Tour early in July of that year, Fokker was in the States developing Atlantic's first commercial plane, the Universal—a hybrid between a large mail plane and a small airliner that might sell under the provisions of the new Air Mail Act. Always quick to grab a business opportunity when he saw one, he immediately realized that the publicity of the Reliability Tour would be just the kind of stimulus his faltering American company needed—if he could put on an exceptional performance. Relying on positive press coverage rather than close working relationships with government bodies had always been at the core of Fokker's marketing strategy, and this was also true of his activities in the U.S.

A combination of ingenuity and a good memory led him to a sure way of achieving an outstanding performance in the Reliability Tour. The idea of the three-engined version of the F.VII had remained dormant since his initial promises to KLM in May 1923. At that time, he was convinced that advances in the reliability of aero engines would soon make multiengined planes obsolete.[29] With a bit of luck, a three-engined variant of the already proven F.VII might just be produced quickly enough to enter the competition on September 26, 1925. The airplane was indeed ready in time, and even performed beyond expectations: Fokker felt an American breakthrough was within reach. The new trimotor would be a double-edged sword, clearing the way for imports of such aircraft from Holland and, at the same time, opening up a market for the American Fokker Universal.

That summer, Fokker's life entered a crucial stage. Going all the way with this new machine, he decided to become a permanent resident of the United States and closely pursue the markets that the Air Mail Act would open up for his trimotor and the Universal. His associates in the U.S., already scouting around for a suitable location in which to expand Atlantic Aircraft, zeroed in on Kansas City, Missouri, right in the geographical center of the country. They hoped to attract one million dollars in capital expansion from Kansas City investors, which would double Atlantic's capitalization—a deal that eventually fell through.[30] After the prototype F.VII-3m was rushed aboard the liner *Veendam* of the Holland-America Line, Fokker waited for the tide before sailing for New York in the late hours of September 8. He must have watched the receding harbor lights with apprehension as the ship put out to sea. In the strictest confidence, lest it would become known in Holland that he was emigrating to the United States, he had, after months of deliberation, finally given instructions for his Amsterdam house to be sold, and his belongings crated and sent to his new home in Alpine-on-Hudson, New Jersey.[31]


## BIG BUSINESS, BIG BUCKS

Arriving in New York six days later, Fokker lost no time restructuring his existing American affairs to ensure smoother operations and be ready for the expected press of increased demand. On September 16, 1925, the Netherlands Aircraft Manufacturing Company was reorganized into a holding company, Fokker Aircraft Corporation, under the combined control of Tony Fokker, Frank Ford, George Davis, and Lorillard Spencer. Soon the Fokker Aircraft Corporation took over all the stock and property of Atlantic as well as the business of coordinating imports from Amsterdam and overseeing all Fokker patent rights in the U.S. Upon its founding, working capital of the Fokker Aircraft Corporation amounted to $1,000,000, but a further capital expansion of $1,590,000, to be raised by issuing new shares, was proposed. Fokker himself was granted a five-year option to increase his ownership in the corporation to 33 percent at a specially reduced rate. Net earnings, the Fokker prospectus promised potential shareholders, would be at least $500,000 annually once the expanded factories had reached their planned production level of a thousand aircraft per year.[32]

With capital expansion thus planned, Fokker considered his American factory well enough poised to enter serious production in the U.S. Moreover, his participation in the Ford Reliability Tour turned out to be every bit the success Fokker had hoped for. Under the direction of his American publicity manager, Harry Bruno, the aircraft was fitted out as a flying billboard, the name Fokker standing out from every angle of view: a conscious attempt to turn the Ford Reliability Tour into the Fokker Publicity Tour. On the list of participants, Fokker's signature was double the size of those of the other entrants. A group picture shows Fokker hogging the limelight, to the visible annoyance of the participant sitting next to him.[33] But the effects of Fokker's primacy in the air tour could not be denied: by October 1925 the American aeronautical scene finally lay at his feet. Orders for his three-engined plane were coming in faster than the small Atlantic outfit could build them, and the trimotor concept itself would dominate international airliner designs well into the 1930s.

That October the prototype of the Universal was rolled out at Hasbrouck Heights. Specifically sized to the emerging U.S. airline market, the four-passenger, single-engined Universal, designed by Bob Noorduyn, represented something of a step backwards from design standards practiced in Amsterdam. While following the customary arrangement of a welded steel fuselage covered with fabric, the Universal heeded the flying practices of U.S. airmail pilots. It had a completely open cockpit, offering its pilots no protection against the elements, whereas in the contemporary F.VII, the pilots were protected beneath the wing and behind a wind-shield. A more dramatic departure lay in the construction of the 47.9-foot wing, for the Universal did not have the wooden cantilever wing Fokker aircraft were famous for. Though the wing was built in one piece, it was thinner, meaning that the main spar did not have the same rigidity as the ones used on the Dutch models. It had to be supported by a pair of steel wing struts attached to the lower fuselage. This unusual arrangement caused the premature end of pilot Nat Brown's attempt to fly non-stop from Seattle to Tokyo in May 1932. To save weight, Brown had the wing struts of his Universal removed and replaced by wires, assuming the struts were superfluous. As a result, his plane crashed into Puget Sound.[34] Nevertheless, the Universal, which could be easily equipped with floats, fit the emerging American and Canadian (bush) market well. From May 1926 to the end of production in April 1931, forty-five were constructed at Hasbrouck Heights, selling at around $15,000 each.[35]

Various approaches to marketing were considered under Harry Bruno's aegis, but it was decided that none were as effective as the publicity brought about by supporting special flights. So when the energetic and well-connected president of Fokker's American launching customer, Juan Trippe of Colonial Air Transport (who led Pan American Airways from 1927 on), approached him to charter the trimotor demonstration airplane for a return flight to Cuba, Fokker was persuaded to come along for the ride. Also on board would be Trippe's millionaire friend from his days in the college flying club of Yale University, John Hambleton, and Hambleton's wife, Peg, whom they picked up in Baltimore en route from New York to Havana. Fokker, by that time more businessman than pilot, did not fly the aircraft himself. Because of his sinus trouble, he wisely decided not to brave the winter cold and stayed in the heated cabin instead, entertaining his fellow passengers and instructing them in the use of the aircraft's toilet—a novel convenience in aircraft of that day. Braving the wintery conditions in the "driving-box" up front were former navy pilot George Pond and Ken Boedecker, a mechanic from the Wright company, which had supplied the aircraft's engines. Boedecker had also accompanied Fokker in the Reliability Tour. The navigator's duty was left to the pilot to protect the privacy of the passengers in the cabin. On the second day of the trip, flying from Baltimore, Maryland, to Augusta, Georgia, the lack of a navigator proved to be a handicap when Pond lost his way and had to put the airplane down in a cotton field to ask for directions.

Peg Hambleton was sitting right in the middle of the tremendous drone of three engines for hours and hours, having to shout to be heard, while the view below offered little more than endless woods and farmland. The first to quit, she deplaned at the next stop, Tampa, Florida, before the five men continued to Miami. There, John Hambleton decided he had also seen and heard enough and left the flight, too. While the aircraft continued to Key West, the last island of the Florida Keys off the coast, an argument developed between Fokker and Boedecker. Because they had not prepared for the flight with any ground organization, obtaining the right kind of aviation-grade, high-octane fuel had become more and more of a problem as the airplane flew further down the coast. Fokker came up with the solution of filling the tanks with a fifty-fifty mixture of regular automobile gasoline and benzol. Though the engines had stood up to this treatment without mishap until the plane touched down in Key West, Boedecker was not prepared to fly on to Cuba, across the open sea, without proper aviation fuel. Fokker accused Boedecker of cowardice, not trusting his own engines; but the mechanic stood up to Fokker's fury, and when the airplane took off for Cuba, only Pond, Fokker, and Trippe were left on board.

Without further problems, they managed to land safely at Campo Colombia, a military training field outside Havana, where they were cordially received by the Cuban president, Gerardo Machado. In warmer climates again, Fokker gave a flying demonstration. Trippe spent his time negotiating with the Cubans, who granted him landing rights for scheduled air services, thus opening an important gateway into the Caribbean that would later prove vital to Pan American.

So far the flight had been a success, but on the way back, trouble struck between Key West and Miami. One engine cut out completely, and the other two were choking on bad fuel. Apparently something had gone wrong at the last refueling stop at a Key West golf course. Unable to maintain the airplane's altitude, Pond was forced to make an emergency landing on what, from the air, looked like a mud flat near Key Largo. He might have known better: the Florida Keys are made of coral. Upon touchdown, the tires were ripped open, and the aircraft came to a sudden and unexpected stop some six hundred feet away from the shore. Shaken, but otherwise uninjured, the three men climbed out into the sea water. Luckily, a local inhabitant came out in a rowboat and put them ashore before they got completely wet. The man in the rowboat went to get assistance but never returned. Like good Boy Scouts, Fokker, Trippe, and Pond lit a fire on the railroad track to stop the next train and hitch a ride to Miami. From there, Fokker sent Boedecker, Pond, and a team of help-ing hands back to Key Largo to change the engines and tires. Pond then took off from an improvised runway of planks and delivered the trimotor to the Fokker plant at Teterboro for extensive repairs.[36] Later, Fokker sold the machine to Byrd's North Pole expedition.

Less than proud of the final episode of this publicity trip, Fokker al-ways kept a low profile on the episode. Despite his confidence in and en-thusiasm for some of the endurance flights with his aircraft, he never again consented to fly in them, preferring to portray himself as "the avi-ator Fokker at his work desk."[37]

From that work desk, so to speak, came the Universal and the subse-quent, larger, six-passenger Super Universal, which was introduced in November 1927. The latter featured a more powerful engine, an enclosed cockpit for two pilots, a fully cantilevered wing, and a modified under-carriage. It went on to become the most successful model Fokker pro-duced in the United States: 137 units were sold. The Universal and Super Universal were the only two models to emerge from Fokker's American production line that could justifiably be claimed to have been fully devel-oped in the U.S. In the context of the large American market, the fact that Atlantic developed only two aircraft was peculiar—evidence of Fokker's powerful hold on his American company. Though he neither controlled the majority of its shares nor held the position of chairman of the board, his reputation and his capacity as technical director allowed him to mon-itor the process of aircraft development closely.

In the field of research and development, Atlantic, as the company was still known then, relied heavily on the work done in Holland. A number of single- and three-engined variants of the Dutch F.VII model were im-ported into the States and assembled at Hasbrouck Heights, with Ameri-can engines and instruments, to be sold as American-built aircraft. These then formed the basis for Fokker's American derivative of the three-engined F.VII, known by its military designation, the Trimotor C-2 (mil-itary type certification was easier to obtain). The main differences were a wider fuselage and the positioning of the cockpit further forward: the air-craft did not have a forward mail compartment like the Dutch models. Some twenty-five were built, most of them for the military, although Pan American Airways also acquired two.

In typical Fokker style, the C-2 was modified into several specially cus-tomized variants, some of which subsequently received separate type des-ignations. Late in 1929 it was marketed as the F-9.[38] And in 1928 it was the starting point for the Fokker Aircraft Corporation's F-10 airliner, which combined a redesigned fuselage and tail unit with a Dutch-built F.VIIB wing. Sixty-five were built, fifty-nine of them the further improved F-10A model, which featured an enlarged, American-built wing.[39] A year before, in the spring of 1927, a prototype of a twin-engined bomber for the U.S. Army had been completed on the basis of the Dutch F.VIII airliner design, using the same wing and fuselage layout.[40]

Such design evolution was typical of Fokker's continued low-cost approach to aircraft construction, also pursued in his Amsterdam plant. Judging by the decreasing amounts of money spent on research and development of the various designs put out by the Dutch factory, stretching the results of the work done in Holland for aircraft development by Atlantic (and subsequently Fokker Aircraft Corporation) must have made Fokker's American enterprise highly profitable. His research and development costs were even lower than in Holland, where design practices were already specifically geared to be inexpensive, enabling the Nederlandsche Vliegtuigenfabriek to recoup its investments with production runs of fewer than ten units. At Atlantic, this approach suited Fokker just fine. It enabled Atlantic/Fokker Aircraft to operate with only a small design staff and thus keep overhead costs low. "Design" work was geared more towards modifying existing designs than producing entirely new ones. Also, with such an approach Fokker's position as chief engineer would not be threatened. As long as essential research and development were done on a tight budget in Amsterdam, he could continue producing low-cost variants in the U.S., directing his American engineers to improvise on proven designs—precisely the sort of work he liked best, and was cut out for.

In the short run, this was a highly profitable approach to aircraft construction. With the bulk of the costs underwritten in Holland, profits could be reaped in the United States with comparatively little investment. In the long run, however, this strategy precluded real development and innovation when Fokker's American competitors were beginning to size up the market and think of ways to produce aircraft that would outperform the Fokkers. In short, Fokker's low-cost successes spurred the competition to come up with something radically new.

With aviation making big headlines in the newspapers throughout much of 1927 as a result of the public's fascination with daring young airmen like Charles Lindbergh, acclaim for Fokker as an aircraft constructor soared too. The success of his airplane in Byrd's polar flight and the nationwide tour of the *Josephine Ford* as a flying billboard for his designs brought much free publicity. Fokker himself became something of a media personality, always keen to talk to the press in his peculiar Dutch accent or speak at gatherings of local chambers of commerce. The high profitability of the Fokker Aircraft Corporation did not go unnoticed either. Though plans for capital expansion in 1925 were not realized, the corporation had not deemed it necessary to issue new shares and continued to operate with its initial one million dollars even after expansion of the Hasbrouck Heights factory. A second plant had been opened in Passaic, five miles from the Teterboro airstrip.

Meanwhile, on the U.S. stock market, aviation shares were rising quickly, and the Fokker Aircraft Corporation was among the first in the undercapitalized aviation industry to attract new venture capital.[41] In November 1927 a group of investors from Wheeling, West Virginia, came up with a package for capital expansion that would, at the same time, resolve the main obstacle Fokker Aircraft had come to face in 1927: an absolute, acute, and insuperable shortage of production capacity to deal with the orders for its airplanes that were now coming in distressingly high numbers, especially for a company that still employed only 266 people in December 1927.

Wheeling had known a short and unsuccessful history in aviation before the offer to the Fokker Aircraft Corporation was extended. In 1920 a local entrepreneur named J. C. McKinley had built a single-engined biplane there at his West Virginia Aircraft Company. But with war surpluses of single-engined biplanes plentiful, McKinley's company folded in 1922. Despite that rapid demise, the aeronautical spirit had evidently lingered in the Wheeling business community, and now a group of local investors combined to form the Ohio Valley Industrial Corporation. It was led by Ralph R. Kitchen, president of a Wheeling construction company of the same name, who offered to invest $3 million in the Fokker Aircraft Corporation if the company was prepared to move production to Wheeling. To make the proposal even more attractive, Ohio Valley Industrial offered to build a manufacturing plant and a purpose-constructed airstrip for Fokker near Wheeling. Kitchen promised to have the new plant ready by April 15, 1928. Badly pressed for extra capacity, Spencer, Ford, and Davis were quick to agree to this lucrative offer, as did Tony Fokker, and on December 2, 1927, the articles of incorporation of Fokker Aircraft were changed to allow the Wheeling investors to come on board. W. P. Wilson, president of the Ohio Valley Industrial Corporation, became the new chairman of the Board of Directors; Kitchen was listed as vice president.[42] To offset Fokker's reduced holding in the corporation, it was decided to drop the name Atlantic altogether, and henceforth use only Fokker, the brand name under which the company's products had been advertised and marketed since 1925.

Construction on the new 800-by-1,200-foot factory, erected at Glen Dale, eight miles south of Wheeling along the Ohio River, began immediately. But despite Kitchen's promises, it was obvious that the new factory could not be ready in such a short time. The official opening of the plant did not take place until August 9, 1928, in the presence of Tony Fokker and four thousand local spectators. The first aircraft, an F-10 for Pan American Airways, was rolled out on December 12. The factory was rented to Fokker Aircraft by Ohio Valley Industrial. Yet of the promised $3 million investment in Fokker Aircraft, the Wheeling investments totaled only $800,000. To make up the difference, a public stock issue was announced on October 2, 1928, valued at $2.2 million,[43] which would allow for rapid further growth on the basis of orders for the Universal, Super Universal, and the various trimotor models.

Fokker Aircraft thus stood to benefit from Wall Street's "aviation boom," which had really taken off since early 1928. Between March 1928 and December 1929, $300 million in aviation securities was floated on the rapidly expanding stock exchange. An extremely profitable market it was, too, for the value of the aviation stock more than tripled in that time to $1 billion, masking the fact that aviation as such ranked only 144th among U.S. industries, and that the value of its products, $90 million, compared rather unfavorably to the total of new investments.[44] Even though aircraft construction was a high-risk business—in Europe, birthplace of the airline industry, none of the air carriers had ever made any money—investors poured into the market for aviation securities. Many believed that air travel would be *the* transportation mode of the not too distant future, and prophesies of the oncoming "Air Age" told of "an airplane in every garage."[45]

The Fokker board hoped to tap into this market to finance the construction of yet another plant, this time in California, to function as a repair facility for West Coast Fokker operators. On October 24, it was announced that the $2.2-million share issue had been successful and that a group of investors headed by the Californian airline Western Air Express (WAE) had acquired all outstanding shares in Fokker Aircraft, thus achieving a controlling interest in the corporation. Western Air Express had been founded on July 13, 1925, under the aegis of James A. Talbot, the president of the Richfield Oil Corporation, to carry mail between San

Francisco (the West Coast terminus of the transcontinental air mail line) and Los Angeles, the headquarters of Richfield Oil. Talbot chose Harris M. (Pop) Hanshue, a former race car driver and car dealer, as the airline's general manager and president. In 1926 WAE landed a contract with the U.S. Post Office for airmail route number 4 (CAM 4), from Los Angeles to Salt Lake City. A very profitable contract it was, too: $1,500 for a single flight, which cost WAE no more than $360 in operating expenses—a return of more than 400 percent.[46] Profits rose even higher when, later that year, Hanshue began selling extra room on his six Douglas M-2 mail planes for carrying passengers. And though WAE lost out to Boeing Air Transport in extending its mail contracts east of Salt Lake City, Hanshue was able to use its strong financial position to take over several smaller operators from 1928 on. The support of Richfield Oil, augmented by a grant from the Daniel Guggenheim Fund for the Promotion of Aeronautics, also enabled WAE to initiate a first-class passenger service between Los Angeles and San Francisco. Hanshue bought three twelve-passenger Fokker F.10 trimotors for the service and began operating on May 26, 1928. Initial results were so promising that, on October 1, 1928, to permit further expansion, Hanshue and Talbot reorganized WAE as Western Air Express, Inc., with a capital of $5,000,000.[47] Three weeks later, the WAE/Richfield combine bought Fokker Aircraft in a strategic maneuver to strengthen its competitive position against Boeing by securing its own aircraft manufacturing operation.

Because Western Air Express was, in turn, controlled by Richfield Oil, that was where the real ownership resided. James Talbot immediately assumed the responsibilities of chairman of the board at Fokker Aircraft and made sure that his right-hand man, Harris Hanshue, became president and general manager at Fokker's, too.[48] As a reflection of the new ownership and to clarify the company's financial situation after the takeover by Richfield/Western, the Fokker Corporation's capital was reorganized. Stockholders were given the opportunity to exchange their shares in the old Fokker Corporation for an equal value of stock in the new Fokker Corporation, and the share-issuing limit was increased to one million shares, valued at $6.6 million. Of that total, 60 percent (600,000 shares) was issued, resulting in a working capital of $4 million, and the remaining 40 percent (400,000 shares) was kept in reserve for future capital expansion.[49] Fokker Aircraft had now really become big business, and with the company thus strengthened, shares shot up 16 percent within a week on Wall Street's curb exchange for unlisted securities.[50]

# 7

## ƦUNNINǴ ON ƐMꝐƬƳ

Tony Fokker's choice of settling in the United States was based on a mixture of business and personal reasons. Holland was definitely too small for him. He felt shackled by the rigors of Dutch society, with its many written and unwritten rules and regulations. America was different: it offered him a chance to do as he liked without such restraints, both personally and in business—or so he felt. As a foreigner in cosmopolitan New York, Fokker could be far more inconspicuous than in Holland, where the press was never far away wherever he went. Late in 1925 he set about changing his American residence from the Waldorf Astoria Hotel in New York to a more permanent home. In May of 1926 he took the logical next step and filed an application to become a U.S. citizen.[1] Following the example of the rich and famous, like the Vanderbilts, to whom he had been introduced in the previous years, he set up a dual residence. He bought a roomy, though somewhat somber-looking modern villa, echoing American design trends inspired by Frank Lloyd Wright, on Hillside Avenue in the sleepy, but not wholly unfashionable hamlet of Alpine-on-Hudson, on the New Jersey bank. In Alpine-on-Hudson he could be close to his factory at the Teterboro airstrip, and yet be assured of privacy. There, on the edge of the Palisades Park, he set up house with his German housekeeper, Christine Döppler, an accomplished cook, who had followed him from Holland.[2] Their attachment did not last long. She left in the spring of 1926 when she noticed that Fokker's attentions were shifting.

One sign of this change was his purchase of an apartment on Manhattan's expensive Riverside Drive to be conveniently near the office of the Fokker Aircraft Corporation on 42nd Street. Strangely contradictory, the peculiar atmosphere of America in the 1920s at once captivated and eluded Fokker. He was not likely to be found in any of New York's theaters or musical establishments, then thriving with the new sounds of jazz; but movies he liked. He abhorred concerts: "the noise bothered him." But he did adopt the American taste for sweets, feasting on them in quantities that showed in his increasing weight and deteriorating teeth. A typical meal, which he was likely to take at any hour of the day or the night, consisted of bouillon, two kinds of pie, four kinds of ice cream, and a desert.

Tony Fokker's personal habits remained as erratic as ever. He hired Helen Kay Schunck, his new personal secretary, on a whim because he liked her looks and the brazen way she refused to be present for work at 7:00 A.M. after staying up until three in the morning taking letters in his room in Detroit's Book-Cadillac Hotel when they first met in November 1925. She found him fascinating, but exhausting to work for: "He keeps one always on the alert as, due to his disconnected method of thinking and expressing himself in English, one must learn to read his mind and anticipate his desires."[3] Convincing Fokker that her services came at a price took several months of extremely long working hours, compensated only by the informal atmosphere. Work was "interspersed with playing the radio, inventing contraptions for the furnace, trying to train 'Bum,' the pet wire-haired fox terrier, in Dutch, German, French, and English; and by servings of hot chocolate and club sandwiches at all hours. Whenever I began to look exhausted, food was ordered, with usually the injunction: 'Now you must write me a memo that I have not starved you.'"[4]

Any place Tony Fokker touched he transformed into chaos, with papers and personal belongings scattered across the floor. Fokker relished his eccentricity. He could be found in his best suit bent over the engine of his favorite Lancia Lambda; or, covered in grease, fast asleep in the back seat of his chauffeured sixteen-cylinder Cadillac, rolling down Fifth Avenue. He would not walk a block if he could find a car.[5]

On the floor of the factory, Fokker was a thorough and exacting supervisor, never quite satisfied with the performance of his subordinates, even though preciously few explicit quality standards were set. But to those who could live with Fokker's erratic behavior, his instant decisions that demanded immediate obedience, he could also endear himself as "Uncle Tony," as his test pilot, Bernt Balchen, liked to call him. A man who would kneel down in a pool of oil to discuss some technical problem with the mechanics working in his plant, he was also a man who would knock on Balchen's door at two o'clock in the morning and take him to the factory immediately to work on design changes when Western Canada Airways had put in a rush order for three Universals after gold had been discovered in northern Ontario in early 1927, and the airline faced immediate demands for additional air transport capacity.[6]

Yet Fokker knew that he was slowly losing his grip on aeronautical technology. To those who asked, he stated defiantly that he had built up no technical library over the years and contemptuously threw out new engineering textbooks as soon as he had glimpsed at them.[7] It was a situation no engineer, self-proclaimed or not, could survive for long, and this was true of Fokker, also. He gradually became more of a figurehead than a technician, playing on goodwill from the past to boost the company's sales. As the 1920s moved to their close, and aircraft constructors started experimenting with new technologies and design methods as big corporate investments kicked in, Fokker's name alone was replacing the quality of his products as the company's greatest asset.

Little by little, Fokker took on the lifestyle of a wealthy businessman. Yet he was a very odd businessman, one who wielded only limited power in the company that bore his name, where the real decisions were being made by the Board of Directors he could not control. These changes were very gradual, but irreversible, nonetheless. Convincing his fellow members of the board, who watched over their considerable investments, that the company was taking the right technical direction became more of a problem. And Fokker himself appeared to be unsure of what direction to take in this respect. Over and over again, he repeated his belief that the U.S. was going to be *the* aeronautical nation of the future, and that small, family airplanes were likely to become just as widespread as cars. He believed that after mass production of airplanes kicked in, their price would go down to around $1,000.[8] It was an argument that had many other supporters. Federal Aviation Administrator Eugene Vidal pursued the idea of a $700 light airplane until 1934.[9] Three different approaches to the development of such small, cheap, and lightweight designs for the private market were initiated by the Fokker Corporation between 1926 and 1928.[10]

At the same time, Fokker was convinced that the future of mass air transport was at hand. Pursuing both scenarios at once, he announced his company had started work on a giant, three-or-four-engined airliner in May 1926. With this machine, for which preliminary design work was being undertaken in Amsterdam, the company wanted to jump into mass transcontinental air transport.[11] Late in October 1928 the blueprints for the new airliner, four-engined in its final guise, were ready. It was to seat no fewer than thirty-two passengers. On the basis of a specification agreed upon by Western Air Express, Fokker's latest financial backer, a contract was drawn up for the sale of six of these machines, designated as F-32's.[12] A month later, on December 2, Tony Fokker proudly announced that construction would begin shortly and that he hoped to roll out the first prototype within four months.[13]

## VIOLET

Early in 1926 Fokker met a young Canadian woman, Violet Austman, in New York. She was a good-looking, twenty-six-year-old brunette, born in Winnipeg on March 4, 1900. An aviation enthusiast herself, Violet was quickly captivated by the renowned, flirtatious aeronautical constructor with his high-pitched voice, thick, funny-sounding Dutch accent, and often boyish manner, who was then at the summit of his success in the United States. The attraction must have been mutual, for after a period of intermittent courtship, they married in July 1927.[14] Like Fokker's first wedding, theirs was a very private affair, conducted in New York before only the witnesses and a few close friends. Not even family members were present. Violet's father—her mother had died earlier—lived 1,800 miles away in Winnipeg, as did her married sister; Tony's mother and sister were both 4,000 miles distant in Holland.

Though they started out as a happy couple, their marital happiness did not last, despite the promise of a fast-moving romance. Perhaps Violet, devoted to what she saw as her wifely duties, followed Tony on more trips and flights than he cared for and got on his nerves. His erratic and often selfish behavior, and uncompromisingly irregular lifestyle, certainly had an effect on hers. Late in 1928 she was admitted to Presbyterian Hospital in New Jersey for what was later described as "a nervous ailment"[15]—a euphemism referring to a condition known in the medical profession of the time as *neurasthenia*: a "profound physical and mental exhaustion."[16] It was used to diagnose people with unexplained chronic fatigue accompanied by such symptoms as nervousness, irritability, anxiety, headache, insomnia, and depression.[17] The standard treatment for these slippery and vague complaints was a number of weeks of absolute rest in the hospital, often with excessive amounts of food, on the assumption that putting on weight would be beneficial to the patient's frail condition. Having put up with Tony Fokker's irregularities for a year and a half, it was not surprising that Violet suffered from neurasthenia-like symptoms. Her stay in the hospital was the prelude to a tragedy that would leave very deep marks on Fokker's life.

Late in the afternoon of February 8, 1929, Violet Fokker came home from the hospital to their Riverside Drive apartment in Manhattan, where the small house staff had been expecting her. She was glad to be home, and although it must have been disappointing that Tony was not there to welcome her, she set about preparing a nice reunion dinner with their cook. As the hours passed, Violet and the housemaid waited in vain. Obviously, yet characteristically, Fokker had forgotten all about her homecoming. When he finally did come in, he was still engrossed in his thoughts after a test flight with one of the new F-10A's, which was proving tail-heavy. Hardly acknowledging her presence, the way he often acted when tired or lost in deliberations, he went straight to the bedroom, lay down on his bed, and fell asleep. To Violet, just back from the hospital, it must have seemed painfully clear that Tony's behavior had not changed in her absence, nor was it likely to ever change in the future. One can picture her pleading with Tony to at least eat something. She had the maid bring his dinner to the bedroom, but there was no waking Tony when he was in this condition. Exasperated, Violet asked the maid to bring her a glass of water while she put on her nightgown. As she stood in wintery cold of the open window waiting for the maid to return, a feeling of utter dejection and loneliness must have descended upon her: living with Tony would always be like this, endlessly waiting, craving for even a moment of his attention—it would never change. The hopelessness swept her clean off her feet. Moments later, her body was discovered on the sidewalk, fifteen stories below.[18]

Upstairs in the apartment, pandemonium broke loose. The maid, returning with the glass of water and finding Violet gone, shook Tony in panic. The window was open, and they instantly grasped the horrible reality of what had just happened. Meanwhile, downstairs, the police had been alerted by the attendants of the apartment building, and when they arrived, they found Fokker ashen and shaking, barely coherent, unable to help the policemen in their inquiries.[19] A physician was called to attend to him, and he remained prostrate the greater part of that and the next day. As for an explanation of what had actually happened, no conclusive verdict was ever made. The original police report stated that Violet's death was a suicide, though this was later softened to "vertigo victim" on the insistence of Fokker's corporate secretary, Herbert Reed, who took care of all press statements about the affair, and who managed to bury public attention within the next two days.[20]

For Fokker, however, dealing with what had occurred proved very difficult and traumatic. Violet was buried in complete privacy in the east addition of the Brookside Cemetery in Englewood, New Jersey, along the road from Alpine-on-Hudson to the Teterboro factory. Afterwards, Tony Fokker had a sculptor make a surprisingly delicate, introverted grave monument—a two-foot-high white marble sculpture depicting a young woman huddled up against a granite rock for shelter. Having failed to provide Violet that shelter in life, this was Tony's way of showing his grief, an attempt to disclose that he had understood and cared for her needs after all. Violet's death would haunt him the rest of his days.

For many hours he must have pondered the course of his life; how it had been shaped; who had been affected by it. The picture that emerged was not entirely flattering: his marriage with Tetta had been a failure that led to divorce; that with Violet had ended in tragedy. His American company was in turmoil, heading for uncertain times, and his Dutch factory was sailing a course of its own. Clearly, reflection was called for, and to aid him in this, Fokker hired Bruce Gould, a young journalist looking for a job, to put his recollections into writing in what would ultimately become Fokker's autobiography, *Flying Dutchman*.[21] With Gould's help, Fokker sought to make up the balance of his life and deal with some of the episodes that had left scars on his memory. Gould followed Fokker in the summer and autumn of 1929, accompanying him on a trip to London, Paris, Amsterdam, and Berlin, but while listening to Fokker's recollections of yesteryear, he was told precious little about Tony's personal life.

When the book finally appeared in April 1931, commemorating Fokker's fortieth birthday, it focused almost exclusively on his endeavors as an aircraft constructor. Perhaps this was because Fokker did not express himself all that easily in English, or possibly because he simply could not open up to Gould about his inner self. But when Fokker read the proofs of the book, he must have felt a sudden, strong urge to explain why his private life had been such a tragic succession of lost loves. For the Dutch translation of the book, he rewrote the chapter on his Russian trip of 1912, relating his short-lived love affair with Ljuba Galanschikova as a prelude to a startlingly honest confession about his relationships with women:

> I have always understood airplanes much better than women. I have had more love affairs in my life, and they ended just like the first one, really, because I thought there could be nothing that was more important than my airplanes. I have always buried myself too deep in my own interests to be able to satisfy the women with whom I was in love. I think too ego-centric. I do not express my feelings, on the naïve assumption that their existence must be understood by intuition. Experience showed me this is false. I have now learned, by bitter experience, that one must give a little too; in love one has to use one's brain just as much as in business, and perhaps even more.[22]

## A HOSTILE TAKEOVER?

In October 1928 the board of the Fokker Aircraft Corporation had unanimously welcomed the capital infusion brought about by the participation of the Richfield/Western group. The deal fit the general pattern of rapidly expanding corporate investment in the American aircraft industry, in which several large groups were emerging beside the Ford Motor Company and the stronghold it had established in 1925 when it bought Stout and started its own airline operation on the side. Those groups were the Curtiss-Wright company, the Aviation Corporation of Delaware (a holding backed by Andrew Mellon), the United Aircraft and Transport Corporation (a coterie of the auto industry and National City Bank, as one author put it),[23] and another group of investors from the senior ranks of the automobile industry called Detroit Aircraft. Small independent constructors such as Grover Loening and the Ryan Aeronautical Company, the builders of Lindbergh's airplane, sold out to these new groups. The larger businesses—Allan and Malcolm Loughead, Glenn Martin, Donald Douglas, and Reuben Fleet's Consolidated Aircraft Company—welcomed the sizable corporate investments with which they transformed their companies but retained a fair degree of managerial independence. William Boeing even continued on an equal footing with United's investors.[24]

With its own shares rising quickly on the stock market, General

Motors stood on the sidelines, watching its big automotive competitor, Ford, strengthening its position in aircraft production with a successful series of all-metal trimotor aircraft. From the point of view of strategic competition, it was inevitable that General Motors would expand into the field of aviation as well. And in view of Ford's increasing trimotor sales, it was equally inevitable that General Motors focused on Ford's main competitor in the field, the Fokker Aircraft Corporation. Besides, Fokker himself had been among those pushing the notion of an aeronautical future in which there would be one private plane for every hundred Americans, a development that, if it came about, could have serious repercussions for the automobile industry.[25] There was some urgency to the matter as well, for the spoils were getting thinner. Early in 1929 Igor Sikorsky sold out to United, which had begun to merge Boeing Airplane and Transport with the other aeronautical companies it held major interests in: Pratt & Whitney (engines), Hamilton Standard (propellers), and Chance Vought (naval aircraft).

The whole aeronautical field seemed adrift in 1928 and early 1929 in a complex series of mergers and stock market takeovers. A sizable number of Fokker shares was floating around too, because some of the earlier investors in the company had sold part of their stock at a profit. Fokker shares were traded with other unlisted securities on Wall Street's curb exchange because the corporation stock was not officially listed in the Dow Jones. On the curb exchange, Fokker Corporation was rapidly becoming the focal point of a high-powered struggle for this last of the leading aeronautical manufacturers that had not yet linked up with any of the big investment groups. In two weeks of increasingly heavy trading, from April 14 to 29, some 50,000 shares worth $1.25 million were acquired by the Universal Aviation Corporation, a latecomer to the market intent on strengthening its position quickly. As a result of this buying, a sudden tenfold increase in trading of Fokker shares occurred on April 22. It was clear that Fokker was up next in the takeover business. In hectic trading on April 22 and 23, when the price of Fokker shares shot up 23 percent overnight, from $33.75 to $41.75, the Aviation Corporation of Delaware decided to acquire Fokker by swallowing Universal. The latter tried to avoid this hostile takeover by pulling as many strings as it could to increase its shareholding in Fokker. If they played it right and acquired a majority in Fokker through fast over-the-counter deals, that might just put Universal's total value beyond Delaware's limits and boost its original Fokker investment at the same time.[26]

In California, James Talbot, Fokker's chairman of the board; and Harris Hanshue looked at this development with mounting concern, because it was unclear what plans Universal, or Delaware, had in store for Fokker if either of them acquired a shareholding majority. So when General Motors president Alfred Sloan Jr. approached Talbot early in May to discuss GM's interest in establishing a shareholding in Fokker Aircraft, Talbot and Hanshue lost no time hammering out an agreement that might be profitable for all parties concerned. On May 16, 1929, after a week of negotiating, it was agreed that GM would acquire the 400,000 shares still held in portfolio at a price of $19.45 per share, which put GM's total investment at $7,782,000, a 40-percent interest in Fokker. Though this was slightly less than half the going stock market price of Fokker shares at the time of GM's offer, it still gave the corporation a handsome extra cash influx of more than $5 million over and above the nominal share value. At the same time, it established GM as the largest shareholder, offered the Fokker Corporation a sound basis for long-term development, and protected the company against a hostile takeover through the stock market, where the price of Fokker shares had risen 60 percent since the first rumors of takeover deal had hit Wall Street on May 10. On May 17, the day after the announcement of the GM acquisition, Universal sold its 50,000 shares, making a profit of $1.9 million on its investment in Fokker the previous month.[27]

Still recovering from his personal loss, Tony Fokker had hardly anything to do with these developments, but they would have dramatic consequences for him in the long term. Again, the top echelon of the Fokker Corporation was reshuffled, this time to make room for GM's representatives. Talbot and Hanshue were maintained in their respective positions, while GM's general manager for exports, W. T. Whalen, joined the board as its new vice chairman, and Eddie Rickenbacker, sales director of GM's Cadillac/La Salle Division and celebrated wartime fighter ace, became Fokker's new vice president for sales. Fokker kept the title of chief engineer. More GM senior appointees followed in July, as GM's grip on Fokker strengthened.[28] On the surface, things appeared to be going all right. The company reported a half-year profit of just under $500,000 and, braced by the GM stock deal, an available cash reserve of $7,592,232.[29] The company even landed a new contract to develop a twin-engined bomber for the U.S. Army (the XO-27) and restated its intention to open up a West Coast facility, at Alhambra Airport, Los Angeles.

Did General Motors know what it was getting into with its 40-percent share in Fokker? It does not seem likely. Though GM's actions originated in the realm of strategic competition, their actual participation in Fokker was probably hastened by the emerging fight for control of the corporation on Wall Street. For GM it was now or never, before the total of required investment exceeded the potential for reasonable returns, because the price of Fokker shares was going up daily. General Motors managed to acquire Fokker at a bargain price, but its board lived to regret it as they gradually found out what they had bought. For beneath the surface, things were not going so well.

It did not take GM's corporate analysts long to determine that a number of things were wrong with the Fokker Corporation. Its long-term prospects were anything but secure, and even less so now, because it was evident that the American and Dutch Fokker companies would grow apart. GM engineers visiting the factory commented on outdated engineering methods, in tune neither with the latest developments in aerodynamic design, such as the Lockheed Vega, nor with the emerging trend towards advanced all-metal construction, as exemplified by the Northrop Alpha. For though Fokker officially held only the modest position of chief engineer, he kept a firm grip on the company through aircraft development. GM's engineers found the Fokker Corporation stepping up airplane size, rather than stepping up technology—a course that Fokker had followed successfully since the appearance of the F.II in 1920. By the summer of 1929, it was already clear that changes were in the air. Bob Noorduyn, Fokker's choice as designer and general manager, was quick to come to terms with these realities and left to join Bellanca Aircraft.

General Motors' top man, Alfred Sloan Jr., decided that some fresh blood was needed at Fokker Aircraft, an infusion of different, ready-made, and proven ideas. To achieve this, General Motors decided to import know-how in metal aircraft construction from Germany. Sloan took it upon himself to travel to Europe and sign an agreement in Friedrichshafen on engineering cooperation with Claude Dornier, founder of the Dornier Metallbauten GmbH. Dornier had a long history in metal-aircraft construction, and that was the way of the future, the GM people reasoned. The dependency on adapting Dutch designs for the American market would have to go. As if to prove the point, the pride of the Dornier Werke, a giant flying boat called Do-X, then by far the largest plane in the world and capable of carrying as many as 169 people, was taken on a transatlantic demonstration tour, eventually mooring in New

York harbor.[30] Significantly, while Fokker was still shipboard en route to
Europe, Sloan and Dornier signed an agreement on October 22, forming
the Dornier Corporation of America under the aegis of General Motors.
It was incorporated in Wilmington, Delaware.[31] When Fokker arrived in
London the next day, he met Sloan there. Already on his way back to the
U.S., the latter informed him that it was GM's firm intention to merge
the Dornier Corporation of America and Fokker Aircraft into GM's new
aviation division. All that would be left for Fokker to discuss were engineering technicalities. Indignant—to say the least—at this blatant show
of power, Fokker boasted to reporters after his meeting with Sloan that it
was *his* intention to weld "his" Dutch and American companies together
as the basis of a worldwide Fokker organization.[32] Nothing came of this
wild suggestion. Instead, Fokker remained in Europe for several months,
reviving old memories and revisiting old places in the company of his biographer, Bruce Gould.

When he returned to New York on March 11, 1930, Fokker came
back to a company that had been through profound changes. Since the
stock market crash on Wall Street's Black Thursday, October 24, 1929,
shares of the Fokker Corporation had dropped 63 percent, and though
the corporation reported a profit of $6,255,402 in 1929, it was a profit
that owed much to the cash infusion by General Motors in May.[33] Shares
in the big automotive corporation itself had fallen 90 percent in value.[34]
In the dramatically changed economic environment of 1930, the Fokker
Corporation now faced the consequences of its continued reliance on increases in airliner size: its new F-32 airliner was far too big and expensive
for the small-scale airlines offering passenger services.


## SEVEN WHITE ELEPHANTS

Tony Fokker may not have been a real engineer, but he was a crafty engineering businessman—be it a short-sighted one. The engineering methods
practiced under his supervision, expanding the size of his proven designs
step by step, were ingenious and inexpensive. But by mid-1926, he was
already starting to lose business to Ford's trimotor designs: all-metal
planes the Fokker Corporation could not build without major investments. A changeover to metal construction required an entirely different
approach to engineering—one that Tony Fokker personally was not
equipped for. If such a change occurred, he would have to give up his po-

sition as chief engineer. To resolve the competitive deadlock, Fokker came up with a truly grand idea: what if they leap-frogged Ford and built a really huge airliner, compensating in size for the lack of applied new technologies? Thus, the F-32 project was born. The F-32 would permit a reduction of air fares on the busiest routes by offering lower operating costs per seat-mile than a conventional ten-to-fifteen-seat airliner. Thus, it was reasoned, the F-32 could be crucial to expanding the market for air travel in the U.S. Preparatory design work progressed steadily, and after the success of the F-10 series, the F-32 was expected to be Fokker's next big breakthrough on the American market. Counting on the F-32's potential for further development, no other airliner projects were initiated.

Preliminary design work on the F-32 started in Holland in 1926. But contrary to earlier practices, most of the labor was carried out in Hasbrouck Heights by Fokker's small design team, headed by Noorduyn and Albert A. Gassner. The latter, a German who had worked for the Austrian division of Albatros during the war, had joined Fokker as deputy chief engineer in 1927. The scale of the project and the size of Fokker's team delayed construction work on the prototype until late in 1928. It was to be a truly grand airplane, with a wingspan of 99 feet and a nearly 70-foot-long fuselage, the nose of which towered more than 12 feet above the ground. The F-32 had a maximum takeoff weight of 22,500 pounds. On September 13, 1929, the F-32 took off on its first fight from Teterboro airport. By then, Fokker had orders for eleven such planes: five for Universal Air Lines and six for Western Air Express. But while Western Air Express supervisor Jim King raved about the luxurious fittings of the aircraft, he also noted an unexpected flaw: "Our biggest problem is to cut the noise, as when you increase the hp, the noise increases considerably."[35]

The noise was caused by the four engines the aircraft needed to fly, hung in tandem pairs under the wing on each side of the cabin. This proved to be a troublesome arrangement, not only because of the noise they created, but also because the rear engines had a distinct tendency to overheat and even blow off their cylinder heads when operated at full power for extended periods of time—a cooling problem that was never resolved. Western Air Express, the only airline to eventually operate the aircraft, therefore flew the F-32 on its Los Angeles–San Francisco route with twenty-eight instead of the intended thirty-two seats. The F-32 did not prove to be the easiest aircraft to fly, either, and pilots reported having trouble keeping it level.[36] Fokker's test pilot, M. Boggs, even refused to take the prototype up under conditions simulating a full operational load.[37]

The prototype F-32 crashed during a test-flight on November 27, 1929, but the most serious problem encountered by the project had nothing to do with the plane's flying characteristics or design as such. With the collapse of the U.S. economy, sales prospects for the big, expensive airliner collapsed too. The F-32 came with a 1930 sticker price of $110,000, more than double that of an F-10A.[38] Despite the explosive growth of U.S. domestic passenger air travel (from 173,000 in 1929 to 417,500 in 1930), none of the airlines dared to risk ordering such a big and expensive aircraft. Indeed, the growth of the American air travel market between 1929 and 1933 resulted from the opening of many new routes, not from a sharp increase in passengers on routes already operated.[39] Only two F-32's were sold, both to Fokker investor Western Air Express. In view of the adverse economic situation, Western was then forced to cancel the rest of their order. Universal canceled all five, and no further orders materialized. Fokker had the last of the seven planes that were built finished as his private, elaborately decorated "air yacht."

The Fokker Corporation suddenly found that it had steered all its resources into what now turned out to be a dead-end street and was unable to recoup the burden of investments in research and development for this giant airliner. With only two units sold, and all foreseeable marketing prospects gone, the seven F-32's were white elephants that had trampled the house of their creator. As a result, Fokker Aircraft plunged into the red. Sloan and the board of General Motors blamed Fokker for making the wrong engineering and marketing decisions.[40] It was the start of a conflict that would drive Fokker out of his American company and very nearly out of aircraft construction altogether, for he had made the fatal mistake of entrusting the company's long-term health to the success of a single airplane.


## GROUNDED

Stepping ashore from the gangplank of the liner *Bremen* of the Nord Deutsche Lloyd on March 11, 1930, Tony Fokker arrived in the United States to witness the demise not only of his F-32, but also of the company upon which he had once built his American dreams. With the market collapsing, General Motors quickly tightened its grip on the Fokker Corporation: GM's shareholding was now approaching 50 percent.[41] Nothing was quite the same as when he had left five months before, and early reorganizations were already resulting in sizable layoffs as production slowed down.[42] Some weeks later, Rickenbacker announced a 20-percent cut in prices on all production models, but to no avail.[43] Further reorganizations and formalization of GM's control as the majority shareholder were now inevitable. At a special general meeting of shareholders on May 25, Harris Hanshue announced that both Fokker and Dornier would be restructured as production companies of a new GM holding: General Aviation Corporation.[44] Tony Fokker could do little else but go along. Three and a half weeks later, at a second special meeting of shareholders, GM dealt the next blow when the majority vote supported a proposal to drop the names Fokker and Dornier from all of General Aviation's products.[45] As a meager compensation, Fokker's position was elevated to that of director of engineering, with a modest annual salary of $50,000 for an initial period of four years, reflecting Fokker's decreased importance to the corporation. His job was remote from the day-to-day running of the engineering department and from direct involvement in aircraft development and construction.

For General Aviation, this shift of power solved only half the problem, and the easiest half at that. Having rejected Fokker, it would have to be rebuilt from scratch. Yet the top-rate, experienced, visionary aeronautical engineers it would need to do so were few. General Aviation's policy between 1930 and 1933 (when the division was sold by GM) reflected this scarcity. First Albert Gassner was promoted to chief engineer while the board pondered further reorganizations. A year later, in July 1931, Herbert V. Thaden, former president of the Thaden Metal Aircraft Company, which had also been taken over by GM in 1929, was brought in. Unable to produce the innovative designs General Aviation needed to survive, Thaden, in turn, was ousted in September 1932 and replaced by Col. Virginius Clark.[46] A graduate of MIT, Clark had been head of the U.S. Signal Corps' engineering department during the war and played a crucial role in updating U.S. military aircraft design. In 1923 he went to work for Consolidated Aircraft, which he left in 1928 to join the American Airplane and Engine Corporation in Farmingdale, New York. There he began developing a single-engined, aerodynamically clean, all-metal airliner that would eventually fly as General Aviation's belated attempt to catch up with innovative design and engineering: the ten-passenger GA.43 of 1933. Only five of these aircraft were built, however. Fokker mediated in the sale of two of them to his old-time European client, Swissair.[47] To emphasize the importance of resolving the engineering deadlock early on, General Motors appointed Jim H. Schoonmaker president of the board in July 1930. Schoonmaker had acquired experience in aircraft engineering at the Dayton-Wright Airplane Company during the war, then went into manufacturing steel springs, and joined General Motors as a consultant at GM's aircraft engineering experimental staff in Detroit in 1929. He replaced Harris Hanshue as president, who now became general manager and was pleased to reduce his General Aviation workload.[48]

Hanshue's Western Air Express had not fared too well in the turbulent year of 1929. As a result of expanding its network beyond routes with post office mail contracts, profitability had taken a downturn, and Hanshue needed a new, lucrative mail route from the post office. In return, Postmaster General Walter Folger Brown demanded that WAE merge with Transcontinental Air Transport and the Pittsburgh Aviation Industries Corporation to form a new and strong airline, then in the process of being incorporated as Transcontinental and Western Air Inc. (TWA) to operate the central transcontinental mail route from New York through Philadelphia, Saint Louis, and Kansas City to Los Angeles.[49]

General Aviation was thus struggling for its survival in an increasingly competitive American aircraft industry, slowing down deliveries of aircraft while inviting new, custom-built designs that could be developed quickly and built cheaply, to keep itself going. At Teterboro a mixed civil and military production run of the F-14 and five search-and-rescue flying boats for the U.S. Coast Guard (the FLB/GA-15) were completed in 1930. At Glen Dale, production of the Super Universal went on until January 1931, while six aircraft of a hybrid version of the U.S. Army C-2 and the F-10A, known as C-7A, were also built there.[50] Things were not going well: General Aviation's losses for 1930 amounted to $2,133,858, and prospects for 1931 looked no better.[51] In the absence of new design projects of real commercial or military potential, the company was sliding toward disaster.

The demise of General Aviation, and of Tony Fokker himself as an aircraft constructor in the United States, was speeded up by an accident near the little village of Bazaar, Kansas, some 100 miles southwest of Kansas City, on March 31, 1931. That Tuesday morning, farmhand Charles McCracken was feeding cattle at about 10:50 A.M. when he saw an airplane come out of the low overcast, going very fast, and almost straight down. One engine seemed to be spluttering, but they all stopped before the plane hit the ground. He could not see where the machine hit the ground but heard the loud crashing noise clearly. Seconds later, a part of the wing came down too. After alarming the authorities by telephone, he waited for the ambulance to arrive from Bazaar, but when they reached the site of the accident, there was nothing they could do: there were no survivors.[52] The accident, tragic though it was for all concerned, would have been quickly forgotten had it not been that one of the six passengers on the TWA F-10A was Notre Dame University's famous football coach, Knute Rockne, a national celebrity, and at the height of his coaching career.[53]

A number of factors came together to make the "Rockne Crash" and the investigations determining its possible cause headline news. President Herbert Hoover called the death of this beloved sports figure "a national loss."[54] It was also recognized that in a declining economy, the effects of a highly publicized crash could have a significant effect on the unfolding U.S. airline industry, thus far continuing to grow against the economic tide. The Aeronautics Branch of the U.S. Department of Commerce, supervising the development of the airlines, therefore decided to break with tradition and inform the public about the cause of the accident as soon as possible.

From the start, the attention of the accident investigators focused on identifying a structural flaw that had caused the aircraft to crash, and with good reason. Concerns about the construction and flying characteristics of the F-10A had a history about as long as the airplane itself. For one thing, the aircraft was tail-heavy and slow to respond to the controls—the very problem that was on Tony Fokker's mind the day Violet committed suicide. After the accident several of its pilots said they were actually afraid to fly the F-10A. Its wings tended to flutter badly in turbulent weather, making the aircraft very difficult to keep under control. But the pilots had remained silent about the problem for fear of losing their jobs.[55]

Some months before the accident, in December 1930, an inspector for National Parks Airways, Dillard Hamilton, had written a letter to the director of air regulation at the Department of Commerce, Gilbert G. Budwig, expressing concern about not being able to inspect the internal structure of the Fokker wings because they were bonded together with glue. Checking the wing spars and the internal bracing would involve removing the plywood covering of the wing and thus damaging it. A month later, the navy tested the F-10A, found it unstable, and rejected it for naval use. Suspicion was riding high even before the crash.[56]

After a few days of making wild guesses about the cause of the crash, investigators focused entirely on the wings of the aircraft, one of which had broken off in flight. Yet Tony Fokker, visiting the crash site in the company of Jack Frye, TWA's vice president for operations, came to a different conclusion. While agreeing that the wing had broken under the effects of what he called "excessive strain," he also noted, "There is no reason to believe that just at this time when weather conditions were unfavorable, a structural failure of the airplane, independent of the weather conditions, could have been the primary cause of the failure of the wing."[57] In defense of his airplane, Fokker observed that inadequate and contradictory weather forecasting by various meteorological stations along the route had been a contributory factor—the pilot should have been warned about local conditions—and he formulated his own conclusions accordingly.[58] But because Fokker's statements did not support the Aeronautics Bureau's investigation into the F-10A wing structure, nor the bureau's wider aim of preventing the crash from becoming an impediment to airline development, they were quickly buried. Indeed, the final report by the bureau's supervising inspector, Leonard Jurden, blamed material fatigue as the most likely cause of the accident. The report said that continuous flexing of the aircraft's wing in over three hundred hours of flying in turbulent weather probably resulted in a spontaneous structural collapse of the wing. Jurden's report also pointed to pilots' observations that the fluttering and flexing of the F-10A's wings was accentuated at precisely the spot where the breakage had occurred. When examining the wing, Jurden also found "peculiar glue conditions, particularly the upper and lower laminated portions of the box spars. Some places the glued joints broke loose very clean, showing no cohesion of the pieces of wood. Other places showed that the glue joints were satisfactory."[59]

Two things set the wings of the F-10A apart from those of the other Fokker models. First, the wingspan of the F-10A was eight feet longer than that of the original F-10 wing (which was imported from Holland and identical to the wing of the Dutch-built F.VIIB). A longer wingspan obviously subjected the outer wing areas to higher stress than in the short-span version. But there was another significant difference: the F-10A wing was the first to be built at the Glen Dale production plant in West Virginia. Bayard Young, a Fokker employee at Glen Dale, hired as an apprentice woodworker in July 1928, described the wing assembly: "The spars were assembled by gluing and nailing ribs between the flanges at specified intervals. . . . Graduated thicknesses of plywood were glued on each face of each spar. They were given final form by planing away the surplus wood with jack planes. A crew of six to ten did this work, working to plus or minus one thirty-second inch which is rather exacting when working with wood."[60] Here, one of Fokker's eternal production problems came up: quality control. The Fokker wings, depending as much on gluing as they did, demanded absolute precision and the highest degree of craftsmanship. Yet the Glen Dale plant was set up to boost local employment and hired woodworkers from the Wheeling area who were unaccustomed to the exacting nature of aircraft construction. The Jurden findings indicated that the one-thirty-second-of-an-inch margin was, perhaps, not always achieved at Glen Dale, which may have explained the "peculiar glue conditions" and the lack of cohesion Jurden found in the wing of the Rockne plane.

On May 4, 1931, after five weeks of deliberations, the director of the Aeronautics Branch, Clarence M. Young, took the unprecedented step of grounding thirty-five F-10A's built in 1929 (at Glen Dale) for passenger service until full inspection of the wings and certain modifications to the ailerons could be completed—a difficult and costly procedure. The airlines were instructed to revise their maintenance procedures for the Fokker planes to ensure against similar accidents in the future (which indeed did not occur).[61]

Tony Fokker was furious that the agency was casting doubt on all of his products because of an accident with one airplane. No accidents similar to the Rockne Crash had ever been documented with Fokker aircraft, though they were in service worldwide and had a good safety record. Storming into Washington, he demanded to attend a meeting between Young and airline officials discussing the ban, but he was not admitted.

Though twenty-five of the affected F-10A's were eventually returned to service, the affair hastened the exit of Fokker airplanes from scheduled service in the U.S. Yet, contrary to what has been maintained in aviation literature,[62] the effects of the F-10A grounding on Fokker's market position in the U.S. were only marginal. Fokker's concentration on the F-32 project, followed by the changes brought about by the GM takeover of the Fokker Corporation and the subsequent forming of General Aviation, had already made Fokker's disappearance as an aircraft constructor in the U.S. inevitable. With the link between the Dutch and American Fokker companies severed, General Aviation had no replacement for the F-10A on the drawing board two and a half years after the airplane's first flight. The F-10A would have been the end of the line for the American Fokkers anyway, because the changeover to metal construction that General

Aviation was pursuing meant starting from scratch. Nevertheless, Fokker felt deeply wronged by the whole affair.

And the ordeal was not yet over. On July 6, 1931, GM appointed Herbert Thaden as the new acting chief engineer and general manager of General Aviation—a move in which Fokker was not consulted, intended to have Thaden take over the responsibilities of both Tony Fokker and Albert Gassner. The appointment was to be confirmed in a general board meeting four days later, on July 10. Here matters came to a head. Fokker had known what was coming and prepared his defense for the occasion. His position at General Aviation had deteriorated continuously ever since GM came on board. Now he had to face the final consequences of General Aviation's decision to abandon the traditional Fokker engineering concepts. Fokker agreed to give up the title of director of engineering and end his direct involvement with General Aviation on the condition that, with his 20-percent shareholding and in accordance with the agreement he had signed with General Motors in May 1930, he would be allowed to maintain his position on the board until the end of his contract in 1934. More important for him, the board also allowed him to regain full manufacturing and marketing rights to, and full use of, the brand name Fokker. The decision meant that the Fokker Aircraft Corporation would be liquidated as a General Aviation subsidiary and that the name Fokker would no longer appear on any General Aviation aircraft. The arrangement suited both parties: Fokker would be allowed to hang on to his former reputation, and General Aviation would cut itself free from the name Fokker and its connection with the Rockne Crash. In a prepared press statement that Fokker and his lawyer, Carter Tiffany, had written before the meeting, Fokker announced the breakup as *his* decision—a move to free himself from General Aviation and found an International Fokker Corporation that would henceforth build and market aircraft of his design.[63] It was never to be. When Tony Fokker stepped out of the meeting on July 10, 1931, his career as an aeronautical constructor in the U.S. was over. In the company of Carter Tiffany and his wife, Anne, who had been personal friends of Fokker ever since they first met in December 1927, he went aboard his yacht *Helga* and sailed away in the setting sunlight.

As a consequence of the split between Fokker and General Aviation, the three Fokker plants were closed in October 1931 and operations moved to the former Curtiss-Caproni plant in Dundalk, Maryland. Financial results continued to be a disaster: a loss of $2,232,735 in 1931 was followed by losses in 1932 and 1933. Between February and April 1933, in extended negotiations, General Aviation merged with North American Aviation, Inc., a holding company with interests in a variety of aviation-related firms that had been formed on December 6, 1928. In the following months, North American expanded further, swallowing up Berliner Joyce. Employment at the former Fokker factory gradually fell from 1,300 to a mere 200 in May 1934. Seven months later, on December 24, 1934, the former General Aviation Manufacturing Corporation became the operational manufacturing division of North American. By that time, Tony Fokker's aircraft construction activities in the U.S. were only a distant echo.[64]

# 8

## WEATHERING THE STORM

### METAL VERSUS WOOD

Throughout the interwar period, the aeronautical community debated the pros and cons of "wooden" aircraft—airplanes built mainly of wood, or of a mixed construction of wood and metal, dominated by wood— versus all-metal airplanes. Though the proponents of wooden construction, Fokker among them, were to be put on the defensive, wood remained the dominant construction material well into the 1930s. In the late 1920s the majority of the wooden planes in commercial operation were Fokker models. Fokker, both in the sense of his companies and the man himself, thus stood at the center of the debate, in which wood became increasingly identified with inertia, and metal with scientific progress—ideas themselves shaped by "the culture of the aeronautical community."[1] Then, and later, the changeover to all-metal construction in airplanes was presented as the inevitable victory of industrial modernity over traditional craftsmanship, an attitude that accepted the superiority of the all-metal airplane as self-evident.[2]

The past chapters have shown that, while such an observation holds elements of validity, it does not reach below the surface of developments. In the case of Fokker, the ideology of metal, and the shift in engineering it brought from art to science, was not the key factor in the decline of the company's position as a worldwide leader in aeronautical technology. In fact, the demise of the wooden airplane in the 1930s had more to do with the failures of the Fokker factories than with any ultimate victory of one technology over the other. That decline itself was brought about by Fokker's reluctance to invest in basic research and development, and by the positions Tony Fokker occupied in Holland and the United States. Whether as owner-director or only as chief engineer, it was he who laid out, or at least had to approve, the course for developing new airplane types. Even had his chief constructor, Reinhold Platz, been willing make allowances for a change of course, declining investments in Holland would have proscribed fundamental innovations in design and technology. In the U.S., poor strategy in managerial decisions—the F-32 project and the inability of General Motors to utilize existing know-how in aircraft of mixed construction—led the Fokker company to lose its position as a market leader in airplane construction between 1929 and 1932. While such processes were closely connected with the person and personality of Tony Fokker, the apparent lack of technological innovation at the Fokker plants also had roots in some of the aerodynamic research conducted outside the Fokker company, the results of which seemed to justify Fokker's choice to continue along the trodden path.

From 1926 on, the National Advisory Committee for Aeronautics (NACA) conducted tests at its Langley Aeronautical Laboratory in Virginia to develop a cowling for air-cooled radial aircraft engines that would reduce the aerodynamic drag such engines caused. By early 1929, the program was showing promising results indicating that the speed of cowling-equipped single-engined aircraft increased 12 to 16 percent— results that were communicated to the industry at once. Subsequently, experimental low-drag cowlings were fitted on a Fokker C-2A trimotor in 1929. To everyone's surprise after the tests with the single-engined planes, the results of the new fittings on the C-2A were extremely disappointing, which appeared to justify Fokker's choice to continue installing engines without cowling on his aircraft. Only after prolonged investigation was the cause of the negative results with the Fokker trimotor discovered: with the two outer engines hanging free underneath the wing, the air flowed back between the wing and the nacelle, disturbing a smooth airflow over the cowling and thus reducing its efficiency. Further tests, confidentially communicated to the army, the navy, and the industry in 1932, brought to light that the optimum placement of an enclosed engine was in the leading edge of the wing, in such a way that the propeller would be well ahead of the wing.[3] Yet because the initial results had been negative, Fokker and his designers were encouraged to continue on their chosen track.

In practice this meant there was little reason for the Amsterdam design bureau to experiment with a different engine placement or with engines installed in enclosed cowlings. Instead, the Fokker engineers adopted the rather less efficient, but cheaper and easier-to-install, British-developed "Townend rings"—a simple metal ring fitted around the cylinder heads to permit a smoother air flow around the engine nacelles. These improved the aerodynamic characteristics of the engines on the F-10A, F.XII, and F.XVIII models. The whole design process remained much the same as it had always been, rooted in practical experience rather than theory and mathematics. For aerodynamic design, the Dutch design bureau depended heavily on the services of the government-run Rijks Studiedienst voor de Luchtvaart (RSL), which was officially only supposed to check the designs to issue certificates of airworthiness.[4] Featuring only minor aerodynamic refinements, the Fokker commercial types generated in Amsterdam between 1928 and 1932, such as the F.IX, F.XII, and F.XVIII, could thus be developed at amazingly low costs. Even the most expensive of the trio, the large F.IX, cost only 29,145 guilders in research and development, less than the going price for one aircraft. The F.XII and its F.XVIII derivative were even more astonishing examples of this practice of shoestring design: total research and development costs for the F.XII were a mere 10,957 guilders; and for the F.XVIII, 794.06 guilders.[5] Such figures indicated that Fokker had become trapped in a downward spiral of development, one in which declining sales precluded changes in investment policy and technology alike. As late as August 1938, the company found itself facing this problem, repeating its argument again and again that the strength and weight of wooden wings were comparable to those of metal constructions, and suggesting that the advantages of duraluminum construction were largely limited to better soundproofing of the aircraft cabin. The real bottleneck in Fokker's continued use of mixed construction was mentioned almost parenthetically: the scale of the investments necessary to produce all-metal aircraft required series production in large numbers, and this was beyond Fokker's diminished sales possibilities. Of necessity, Fokker had become specialized in building small numbers of highly customized aircraft.[6]

Long before that, low investments had made it obvious that the Dutch Fokker company was no longer keeping abreast of contemporary trends in aeronautical engineering and could only be heading for trouble. That they were able to turn out reasonably successful new aircraft types at all was little short of miraculous and served to credit the management and staff Fokker had hired. Bruno Stephan and his team were operating in a continuous predicament. Their authority to run the company was strictly limited because all major financial decisions required the prior approval of Fokker himself.

In the airline business, the lack of innovation at the Fokker plant was already being watched with rising apprehension. In a letter to the board of KLM, Albert Plesman complained bitterly about this because the airline was seeking to introduce more efficient aircraft that would cut its costs per ton-mile (the standard airline unit of production).[7] One of KLM's main financial backers, Frits Fentener van Vlissingen, who also occupied a seat on the Board of Control of the Nederlandsche Vliegtuigenfabriek, tried to persuade Fokker to accept a large capital injection into his Dutch company, breaking Tony Fokker's grip on the ownership of the factory and changing it into a publicly owned company. Such a change would make larger resources available for stepping up much-needed modernization. But Tony Fokker, already fighting the consequences of large corporate investments in the U.S., would hear none of it.[8] In September 1932, KLM provided the company with a direct incentive for innovation: a contract to modify one of KLM's twin-engined F.VIII's by incorporating engines placed in NACA cowlings in the wing's leading edge.[9] While aeronautical construction in general was striding along technology's natural path of development—toward increasing complexity—the Fokker company had to be dragged into making innovations by its own customers.

The situation across the Atlantic was quite different. In the U.S., the rapid expansion of commercial air transport after 1929 produced a sense of competition between the leading airlines far stronger than in the government-subsidized environment of air transport in Europe, where all international services were generated under fifty-fifty revenue-sharing agreements called "pools." In a stagnant economy, operating the most economical equipment available was vitally important to cutting operating costs. The American aircraft industry had to keep on its toes, and producers watched each other's advances sharply, quick to copy new ideas that worked. The results of the various NACA research programs, communicated to the industry through NACA's annual aircraft engineering conferences, provided further incentives for innovation.[10]

The accumulated effect of these factors was that America's leading constructors overtook the Europeans in aircraft design in the early 1930s.
Following the trend set by Jack Northrop's revolutionary Lockheed Vega
of 1927, American aeronautical engineers were experimenting with new
approaches to design: higher wing loadings, monocoque fuselages, aerodynamically cleaned-up airframes, controllable-pitch propellers, and retractable landing gears. The Vega served as a model that was followed in
Reuben H. Fleet's Consolidated Fleetster of 1929, developed for Pan
American Airways' competitor, the New York Rio and Buenos Aires Line
(NYRBA), of which Fleet was one of the prime movers. Giuseppe Bellanca's designs of single-engined sesquiplanes like the P-100 Airbus of
1930 incorporated some of the same aerodynamic refinements. That
same year, Boeing presented its sleek, low-wing, all-metal Monomail; and
Jack Northrop his commercially more successful Alpha, the first airplane
to combine a monocoque fuselage with a multicellular system of wing
bracing that reduced the dangers of wing flutter, and wing flaps to reduce
landing speed. Lockheed culminated its own rapid succession of radically
streamlined, stressed-skin aircraft in its Model 9, Orion. Three years
later, in 1933 and 1934, three even more innovative airplanes entered the
market as a result of these developments: the Boeing 247, the Lockheed
L-10 Electra, and the Douglas DC-2.

The Boeing 247 was the first product of this design trend to become
available for commercial airline use. It made its first flight on February 8,
1933, and Boeing's associated company, United Air Lines, became its
first, and initially exclusive, operator. As fast as any fighter aircraft of the
day, United's sleek, shining, low-wing ten-seater offered substantial savings in operational costs as well. It was the immediate envy of the airline
business, and United's competitors stood in line to buy some. But Boeing
claimed that it was under contract to deliver the first sixty production
247's to United. Jack Frye of TWA, United's biggest competitor, therefore
wrote a letter to a number of producers enticing them to come up with an
airplane of even better size and performance. TWA's contract, for which
General Aviation competed in vain, was eventually won by a revolutionary design—the Douglas DC-1—produced by a group of highly educated
and skilled engineers in and around the Douglas company of Santa Monica, California. By a series of lucky coincidences, Donald Douglas, who
had, thus far, built only sturdy army aircraft and was hoping to diversify
his product range, assembled a design team that brought together Harvard and Cal Tech engineering degrees with the proven design talents of
Jack K. Northrop.

Fueled by a competitive ambition to come up with an airplane that would outclass the 247, Douglas did just that. Though heavier than specified, and with two rather than the three engines Frye had asked for, the DC-1 carried twelve passengers to the 247's ten and was 15 miles-per-hour faster than the Boeing rival. It first flew on July 1, 1933, and TWA was so pleased with its initial performance results that it ordered a batch of twenty-five of a slightly stretched version, the DC-2, seating fourteen passengers. These were very expensive aircraft to develop and absorbed a total of around $307,000.[11] Douglas could only afford to develop them because he was lucky enough to head the first U.S. aircraft manufacturing company to secure revolving bank credit. He then took an enormous gamble with the DC-2 in putting its unit price at only $65,000, which meant he would have to build at least seventy-five of them to break even. It proved to be a sound strategy: the modest price of the DC-2 encouraged airlines to accept the standardized design concept of the DC-2 rather than commission their own highly customized aircraft somewhere else. Besides, the DC-2 offered operational savings over every other airliner then in use. Douglas sold 190 of them before production was discontinued in favor of the DC-3.[12]

## DISAPPEARING MARKETS

These developments would have a serious and lasting impact in Europe, where Fokker's Nederlandsche Vliegtuigenfabriek found itself in a predicament after 1929. The export markets, on which it relied, were quickly drying up as a result of the world economic crisis and the gaining momentum of the international disarmament movement through the League of Nations. While the former cut directly into the civil markets, the combined effects of economic crisis and disarmament reduced the volume of military orders. Besides, the Fokker management was consciously steering clear of all financial risks.

In the autumn of 1930, the Fokker D.XVI fighter won a Romanian Air Force competition for a new first-line aircraft to replace the aging D.XI's that Romania had acquired in 1924. This opened prospects for orders of fifty to one hundred new aircraft. In Bucharest, Fokker's sales representative, F. Rasch, reached an agreement for an initial batch of fifty airplanes on the condition that his superiors would agree to the Romanian demand to defer payments up to five years. Both Fokker and Stephan decided that the risks of such a contract were too high because the Romanian currency, the leu, was under heavy pressure on the foreign exchange market.[13] Negotiations were deliberately perpetuated without any serious attempt to arrive at more acceptable conditions of payment, in hopes that Romanian interest in the D.XVI would eventually wane without having to affront the government by doubting the value of its currency. Such a development indeed occurred when the Romanian crown prince, Nicolas, visited Poland in September 1931 and was given a full red-carpet treatment at the Państwowe Zakłady Lotnicze (PZL) aircraft factory. Though the PZL P-7 had come in last in the 1930 competition, Prince Nicolas came back enthused about the improved version of the P-7, the P-11. Much to the displeasure of the Dutch envoy in Bucharest, who had mediated in the negotiations, the Nederlandsche Vliegtuigenfabriek happily let the order slip, and PZL delivered a series of fifty P-11b fighters to Romania in 1934.[14]

Similar developments occurred around a possible order for Fokker aircraft by the Chinese Kuomintang government in Nanking. As part of a sales tour through Siam, China, and Japan, Fokker's demonstration pilot, George Sandberg, gave a series of flying displays with a D.XVI in Mukden, Manchuria, early in 1931. The Chinese were impressed. Rasch subsequently managed to hammer out a contract for thirty D.XVI's and thirty C.VE's with the "Young Marshall," Chang Hsüeh-liang, who controlled most of Manchuria and northern China at that time. Yet here again, Rasch made promises the company's management was unwilling to fulfill. After the provisional contract was signed, he received instructions from Amsterdam to specify the conditions of payment. Marshal Hsüeh-liang was deeply insulted and enraged. Before the Netherlands' diplomatic envoy, Willem Oudendijk, he cried out three times, "They tried to fool me!"[15]

Now eager to back out of the contract, Hsüeh-liang demanded a second series of demonstration flights, to be held in Peking. Rasch and Sandberg were put in a predicament: if Amsterdam was at best lukewarm about the order, should they proceed with their efforts? And if not, how would the Chinese then react? Would Hsüeh-liang even allow them to leave after insulting him twice? On September 18, 1931—the very day Japanese troops began their invasion of Manchuria to overrun Hsüeh-liang's forces and create their petty vassal state, Manchukuo—Sandberg was told to take his D.XVI to 6,000 feet and make a vertical dive with the engine at maximum power. Sandberg, judging this maneuver to be extremely hazardous, refused—reason enough for the Chinese to cancel the order a month later despite the ongoing conflict in Manchuria. In Amsterdam, nobody felt sorry, even though they presented the Kuomintang government with a hefty claim for breach of contract.[16]

For the Nederlandsche Vliegtuigenfabriek, the consequences of shunning such contracts, risky though they were, would be considerable. The company's prominence in the international military field was lost between 1930 and 1932. Without further export orders, production collapsed, and layoffs increased dramatically. In September 1932 Plesman was but one of a number of worried people in Dutch aeronautical circles who feared that the company would shortly go out of business. To prevent this, discussions were initiated with Fokker on a future replacement for the F.XII's and F.XVIII's that KLM was using on its East Indies route. At KLM, the thinking was directed toward a large, possibly four-engined airliner to be delivered in 1934.[17]

In the meantime, it was imperative that the Fokker factory remain open, which the government also considered to be in the national interest. To alleviate immediate needs for production orders, the Air Corps commissioned a series of twelve D.XVII fighters for no other reason than to keep Fokker in business.[18] The order came just in time to save the company from final collapse, for the Nederlandsche Vliegtuigenfabriek had plunged into the red in 1930 and faced mounting losses despite a total of 923,000 guilders that was withdrawn from the company's reserves between 1929 and 1934 to offset some of the unexpected setbacks. Before the first signs of recovery in 1936, the company continued its downward slide, suffering a record loss of 3.5 million guilders in 1935.[19] Employment fell from around 600 in 1929 to only 130 (68 in managerial and staff positions, 62 workers on the factory floor) in December 1932.[20]

With the company reduced to such a small size, comeback efforts concentrated on the field of civil aviation. Even here, the task ahead was daunting, and the objectives were far from clear. From his American debacle, Tony Fokker had brought back more than a deep-rooted fear of corporate investment. He also diagnosed perceptively what had gone wrong with aircraft development by the Fokker Aircraft Corporation of America: it had concentrated all its resources in developing an airliner that was much larger than the airlines needed at the time (and hence was too expensive) instead of stimulating air transport growth by providing more services. After being ousted by General Motors, he must have resolved not to make the same mistake again. Indeed, the growth in numbers of airline passengers in Europe between 1930 and 1939 followed the same pattern as in the U.S., resulting in 68.5 percent more flights, not in a spectacular increase in aircraft size.[21]

Fokker realized he must build not a bigger plane but a more economical one. Uncertain whether or not he could keep the company going with the heavy losses it had experienced, he saw to it that work started early in 1932 on a follow-up to his successful European transport aircraft, the F.VII and F.VIIB-3m. Planned to be ready in April 1933, the new airplane, the F.XX, sat twelve passengers and incorporated such novelties as enhanced streamlining and a retractable landing gear. It also made use of the NACA research on engine cowlings. By pairing these innovations with the established Fokker construction methods, research and development costs could be kept down to an affordable level, despite the financial hardships the company was going through. The resulting F.XX cost a modest 80,197 guilders ($48,600) to develop, a far cry from what Douglas was spending on the DC-1 and DC-2.[22] Fokker had high hopes that it would hit the market as exactly the right airplane at exactly the right time.

It was not to be. Several factors impaired the success of the F.XX project right from the start. On the technical side was the assumption that the innovations would produce a much more economical plane. But the new engine cowlings did little to improve the aerodynamic characteristics of engines hung under the wing, and the latest results from further NACA testing became available only when the F.XX had already been designed. Also, a retractable undercarriage, especially in an aircraft of high-wing configuration such as the F.XX, entailed substantial weight increases and thus higher fuel consumption. There were sound financial reasons for incorporating many of the established engineering practices established in past Fokker planes in the F.XX, but the end result was an aircraft that did not offer the improved operating economics that could have made it a success.

On the marketing side, the problems were even more serious. In the past, Fokker had generally developed aircraft on the assumption that a good airplane sold itself. Once a new airplane proved satisfactory, the airlines would line up to buy it. Apart from the F-32 mishap, this system had worked well, and Tony Fokker felt confident it would work with the F.XX, too. Unfortunately, the F.XX lacked an enthusiastic launching customer. Unimpressed with what Plesman considered only minor operational savings, KLM was dragged into buying the F.XX only after long and difficult negotiations, when Fokker finally offered a 20-percent rebate. The F.XX was taken over by the airline on November 20, 1933, for 60,000 guilders.[23] The difficulties of negotiating the sale stemmed from the long history of personal animosity between Plesman and Fokker, and, more important, because Fokker proposed the F.XX as a replacement for KLM's older generation of Fokkers on its Amsterdam-Batavia service. Yet at that time, midterm planning at KLM for developing its Indies route focused exclusively on a much bigger airplane, which Fokker had been commissioned to build: project Fy, later named the F.XXXVI.

This placed Fokker in a no-win situation: while the F.XX design might have been improved upon, resulting in a good second generation of Fokker trimotored aircraft, he was faced, on the one hand, with an unwilling customer bent on developing a larger airplane, and, on the other hand, a prototype F.XX that turned out to be less economical than hoped for. The only other prospective buyer for the F.XX, the French carrier Air Orient, canceled its order for three F.XX's in January 1934, and in the end, only one F.XX was ever built.[24] Production orders for the Fy project, meanwhile, seemed fairly certain to emerge, and with the Nederlandsche Vliegtuigenfabriek in rapid decline in 1932, there was little Tony Fokker could do but agree to develop the larger aircraft, even though he did so against his better judgment and openly communicated his lack of enthusiasm in his contacts with KLM.[25] It was obvious that the long-standing link between KLM and Fokker—never very secure—was about to break. Only with great difficulty could the airline be persuaded to buy Fokker's latest product; and only with great difficulty could Fokker be persuaded to develop a machine to the airline's latest specifications.

Subsequent events proved that Fokker was right about the prospects for the F.XXXVI. The 296,900 guilders spent on research and development for this huge aircraft, which could seat up to thirty-six passengers or accommodate sixteen sleeping berths on intercontinental routes, bore no relation to its sales: the one machine built was sold to KLM for 120,000 guilders. Fearing the aircraft would turn out to be far too big for the European air transport market, and determined not to put all his eggs in one basket, as he had with the F-32, Fokker decided to develop simultaneously a down-scaled variant, the F.XXII (seating capacity twenty-two) to be used on Europe's air routes. It cost the company an additional 304,000 guilders. Four F.XXII's were sold: three to KLM, and one to the Swedish carrier Aktie Bolaget Aerotransport (ABA).[26]

After an agreement was reached with KLM in November 1932 for developing the F.XXXVI, the design was completed fairly quickly. In keeping with the latest NACA findings, the F.XXXVI and F.XXII now incorporated engines built into the leading edge of the wings and enclosed by streamlined cowlings. They also had smooth joints between the wing and the fuselage for further drag reduction. Nevertheless, the Fokker design team followed tradition in both aircraft types, once again choosing a mixed construction of a linen- and wood-covered, steel-tubed fuselage and a wooden wing in a high-wing layout. The high wing created a serious problem for the construction of the landing gear, and with the experience of the F.XX in mind, the engineering challenge was solved by a permanently fixed undercarriage. The giant, man-size wheels protruding beneath and at the sides of the plane's fuselage were an ugly, if lightweight, solution that damaged the attempt to create an image of modernity by painting the aircraft with aluminum dope. The cockpit layout was Tony Fokker's own design: the copilot sat about two feet behind and to the right of the pilot, instead of the customary side-by-side arrangement. He promoted the design as a way to improve the pilot's all-around view, but it was more likely the result of constrained cockpit space in the streamlined nose of the aircraft. Such features made the F.XXXVI and F.XXII designs odd mixtures of old-fashioned 1920s and emerging 1930s engineering trends.

Together, the sheer size of the new aircraft—the F.XXXVI had a wingspan of just over 108 feet and a length of 77 feet— and the technical innovations that were introduced with it posed tough, time-consuming engineering problems, and the project was seriously delayed. Originally planned for late April 1933, the first flight of the F.XXXVI did not take place until June 22, 1934. In the subsequent nine-month test period before the aircraft was ready to go into KLM service with KLM in March 1935, seventy-three faults had to be corrected, and in its first six months of airline service, the airplane was grounded for eighty-six days because of malfunctions. Engine-cooling problems limited the F.XXXVI to cruising at 56-percent engine power.[27]

That long period of development was fatal for the aircraft's marketing prospects and for Fokker's plans to reestablish the Nederlandsche Vliegtuigenfabriek as a leading producer of civil airliners. With the F.XXXVI/F.XXII projects, Fokker had fought an uphill battle. From the start he had been skeptical about the F.XXXVI's possibilities, but with KLM's virtual rejection of the F.XX against a pending order for six F.XXXVI's, the development of the aircraft could hardly be discontinued. In 1933 and

1934, crucial years in the history of aeronautical engineering, the F.XXXVI and F.XXII projects soaked up money at an unprecedented rate, blocking any possibility that Fokker might hook on to the process of technological change that was going on in aircraft construction in the United States. It was a cruel irony that because of Plesman's insistence on the F.XXXVI project, Fokker made the very same mistake in his Dutch enterprise that had cost him his leading position in the U.S. When the F.XXXVI was finally taken over by KLM on March 16, 1935, the technological environment of air transport had changed dramatically, and KLM's midterm planning for the Indies route had changed with it. Once again, Fokker found himself without an airliner that fit the needs of the market.

## DEALING WITH DOUGLAS

American developments in aeronautical construction would have made their impact across the Atlantic with or without Tony Fokker. But as it was, they directly influenced his life and the fortunes of his Nederlandsche Vliegtuigenfabriek because of decisions made in the boardroom of KLM in November 1933. A month before, early in October, Plesman had sent one of his pilots, Koene Parmentier, an energetic captain who was rapidly being promoted through the ranks, on a fact-finding mission to the States to study the latest American approaches toward night flying. In the course of his travels, Parmentier was given the opportunity to check out TWA's brand new DC-1 at the Douglas plant in Santa Monica. He was immediately enthused about its performance. Indeed, he was so impressed that he wrote a special report on the characteristics of this machine and its successor, the DC-2, then under development.

The report hit KLM's management like a bomb, completely upsetting aircraft procurement planning as the projected technical characteristics of the DC-2 sank in. Here was a machine that outperformed all other commercial transports in every way. Furthermore, the Parmentier report arrived at KLM's head office in The Hague when the airline was becoming increasingly worried about the setbacks Fokker was encountering in the development of the F.XXXVI. With traffic on its Amsterdam-Batavia service expanding more rapidly than anticipated, the airline was anxious that the six F.XXXVI's might not be ready in time to meet the increasing demand for seats on its prestigious route. Behind closed doors, Plesman and his chief engineer, Pieter Guilonard, had criticized what they saw as a lack of technological innovation at the Fokker plant for several years, seriously doubting that the Nederlandsche Vliegtuigenfabriek could survive in the long term unless Tony Fokker set about modernizing soon. Besides, personal antagonism between Plesman and Fokker was running high again—this time about the way Plesman felt Fokker was pressuring KLM into buying the F.XX, for which the airline had lukewarm feelings at best. These factors led Plesman to make a strong plea to his Board of Directors on November 28, 1933, to be allowed to order one DC-2 so that KLM could evaluate it in operation. Careful not to offend the patriotic feelings of his board members, he proposed a study of the DC-2's suitability as a stopgap airplane to meet the demand on the Amsterdam-Batavia route until the F.XXXVI's became available.[28] Plesman's carefully worded proposal reflected a sudden and fundamental change of mind: instead of meeting the increased demand for seats with a bigger aircraft, KLM would now see if it could meet that trend by increasing the frequency of its service. When his board consented, two key problems remained to be solved: how to go about ordering the plane (KLM had no contacts in the U.S.), and how to obtain the necessary approval from the government, which held a controlling interest in the airline. Plesman had his answer ready: involve Fokker in the deal.

That very evening, he invited Fokker to dinner at his home, a rare occasion of informality between the two men. Inevitably, the subject of conversation turned to the Douglas order, and Plesman indicated that he would like Fokker to act as a middleman between KLM and Douglas. Fokker immediately realized the wider implications of what Plesman was suggesting. Shaking Plesman's hand in agreement, he jumped up from the table, grabbed what loose change he could find in his pockets, scattered it on the dining room floor before Plesman and his bewildered children, and left in a hurry.[29] Wiring ahead, Fokker set sail for New York that very evening, leaving a satisfied Plesman to inform the Ministry of Waterworks on the next day that it pleased him to see Fokker so "cooperative" in acquiring the most modern and economical equipment for KLM.[30]

After initial negotiations with Donald Douglas in California, Fokker and Douglas signed an agreement on January 15, 1934, the essential clause of which (article 5) read: "Douglas . . . hereby appoints Fokker as his sole agent in Europe, with the exception of Russia."[31] It was a mutually satisfactory agreement. For Douglas, it meant expanding the market for the DC-2 across the Atlantic at no additional cost. For Fokker, it meant securing control over the marketing of aircraft of a suddenly important competitor. He personally bought ten DC-2's "off the shelf" from Douglas at prices ranging from $71,700 to $74,265 each (depending on the finishing of the planes), to be delivered to him with priority between June 15 and December 8, 1934. For a further $100,000, Fokker also bought all rights to license-manufacture the DC-2 and future twin-engined derivatives in his Amsterdam factory for a period of five years.[32]

But was he really planning on license-manufacturing the DC-2 in Amsterdam? Although the Douglas contract would provide cheap know-how about the latest technology in the production of all-metal stressed-skin aircraft, reequipping the factory for such production would require major investment and relinquishing his hold on the engineering process. Fokker fully realized the importance of the Douglas's technological achievement with the DC-2, but he was not at all sure if his company would be able to catch up quickly enough to recoup part of the market before the name Fokker would be history. However, by acquiring the rights for license production, he at least ensured that none of his European competitors would get ahead of the Nederlandsche Vliegtuigenfabriek while he made up his mind about its future. To shore up the position of his Dutch company further, he also bought the sales rights to the new Lockheed L-10 Electra in February 1934.[33]

If the acquisition of the DC-2 started in a spirit of cooperation between KLM and Fokker, such concurrence did not last long. Problems started before January 1934 was out, when it became evident that the construction of the DC-2's tail surfaces did not meet the Dutch requirements for a certificate of airworthiness. Without such a certificate, KLM would not be able to use the aircraft. The question was: who would have to foot the bill for the necessary modifications? The matter immediately grew into a serious conflict between Plesman and Fokker, when Fokker suggested that if KLM did not want to pay for the modifications, he could always withdraw from the deal altogether, close down his Dutch plant, and spend his time the way he pleased. An enraged Plesman aired his feelings, without mincing words, a month later: "And now it is a disgrace how someone who is half German, half Dutch, half American, juggles about with our interests. The aviation industry needs to be pinned to Dutch soil. It should not be possible that mister Fokker goes off to America and simply 'closes the place down,' like he says."[34] A solution was not reached until August 1934, when the Dutch authorities agreed to issue a certificate of equalization upon presentation of a U.S. Air Transport Certificate, which temporarily settled the matter. Shortly afterwards, on September 12, KLM's DC-2 was unloaded on the Rotterdam quayside.

By then, much had changed. First, KLM had completely abandoned its plans to acquire F.XXXVI's for its East Indies route. Calculations had shown that two flights a week with DC-2's would solve its seating shortage. Second, the nine DC-2's the airline needed to operate such a service would cost less than the six F.XXXVI's it had planned to buy originally. Besides, the DC-2's were faster and about 15 percent more economical in fuel costs. As soon as the matter of the airworthiness certificate was settled, Plesman was authorized to approach Fokker about an order for fourteen more DC-2's—eight for the East Indies route, and six for KLM's expanding European network.[35]

On August 14, negotiations between Plesman and Fokker on the largest airplane order in KLM's history up to that point began in an atmosphere of anxiety and mutual distrust. The price of the DC-2's was the pivotal issue. Tony Fokker, smugly waving his exclusive sales rights in Plesman's face, offered the DC-2 at a base price of 160,000 guilders (about $97,000 in 1934 dollars). Plesman nearly had a fit. The meeting broke up in the worst spirit imaginable, and Tony Fokker decided he would take some days off to catch his breath.[36] Four days later, Plesman came with a counteroffer: a fixed price of 125,000 guilders ($76,000), which, KLM had calculated, would still leave Fokker a margin of about 10 percent on each aircraft. Lenience on the price of the DC-2 was not, of course, in Fokker's interest, especially since KLM's move to DC-2's was likely to seal the casket on civil aircraft sales by his own company. Yet he suggested that arbitration might solve the problem. Fearing that arbitration might force KLM to accept a median price far above what he was offering now, Plesman refused to discuss the idea. He and Fokker drifted farther and farther apart the longer they negotiated. On the evening of August 22, the two men had yet another long and uncomfortable standoff. Fokker refused to budge on the price issue. With barely concealed anger, Plesman wrote to Fokker the following day: "You must understand very well, that this way of doing business—the way You suggest—is not what it is supposed to be, and that it will be extremely difficult to convince our Board of Directors that the way you wish to treat this order is acceptable."[37]

Fokker remained unperturbed. He knew this was the end of the road for his company. In the months leading up to the negotiations, Plesman had conducted a campaign of vituperation against Fokker and his company in government circles to beef up support for KLM's changeover to Douglas. Before the Van Doorninck Committee, on the future of the Dutch aircraft industry, Plesman was taken down as saying, "Finally I have to tell you that Fokker has a factory that is worthless. I warned him a long time ago. Everything has to be transported by boat from the ELTA, and disassembled to be brought to Schiphol. The factory is also much too small. The machines cannot be properly placed, or turned about in it. It looks like a dump. The factory should be located next to a big airfield. Fokker's building is old trash."[38]

Having gone this far, Plesman would certainly do everything in his power to avoid having to buy Fokker aircraft ever again. And for his part, Fokker decided never to fly KLM if he could possibly help it. Henceforth, his airline of choice was the Deutsche Luft Hansa. Hardly an issue was left about which the two men were on speaking terms. And then Plesman made a fatal mistake. Propelled by his anger, feeling certain he would prevail against Fokker in the end, and anxious to get rid of KLM's old planes, he sold off eight of KLM's Fokker airliners in September (an F.VIII to ABA and five F.XII's and two F.XVIII's to the Czech airline CLS). But because delivery dates were written in the contracts, Plesman had also sold away his room for bargaining. His haste would cost KLM dearly.[39]

In the early morning hours of October 20, 1934, twenty-one airplanes took off one by one from the Mildenhall air base in Suffolk, England, in the greatest aerial derby in the world: the England-Australia Race. Organized by a committee chaired by Sir MacPherson Robertson, a wealthy Australian industrialist in confectionery, it was intended as a grand celebration of Melbourne's centenary. Convinced of the great potential of the aircraft, Plesman had entered his brand-new Douglas DC-2 in the race, registered PH-AJU, and named *Uiver* after a subspecies of storks. In doing so, KLM was pursuing an agenda well beyond that of the aerial derby itself. Since 1929 Holland, Australia, and Britain had held intermittent, and entirely unsuccessful, negotiations about Australian landing rights for a Dutch airmail service to Sydney. Frustration over the continued Anglo-Australian rebuff of the Dutch request ran high in The Hague. This was KLM's chance to demonstrate how fast and efficient such a service could be compared to Britain's Imperial Airways flights, which took fifteen days.[40] Parmentier, the pilot who had first called attention to the DC-2, was granted the honor of piloting the plane. With his copilot, Jan Moll, and crew members Bouwe Prins and Cornelis van Brugge, he flew the *Uiver* to international fame and fortune, arriving in Melbourne three days later, second only to the purpose-built British de Havilland DH-88 Comet.[41] The feat brought KLM considerable international acclaim and did much for overseas sales of the DC-2. But in the short term, at least, it did KLM more damage than good. The success of the *Uiver* solidified British and Australian prestige-inspired resistance to granting mail rights and tied KLM even more tightly to acquiring more DC-2's. On October 30, KLM's board authorized Plesman to continue discussions with Fokker, now to a maximum price of 176,000 guilders ($107,000) per aircraft,[42] 10 percent over Fokker's asking price.

Indeed, in the new round of talks between the two parties that followed, the price could only go up. It also gradually became clear that Fokker was stalling the negotiations on the basis of various technicalities and delivery charges. He had plenty of time, because he had no intention whatsoever of reequipping his factory for license production of the DC-2, having long since decided to import the aircraft directly from the U.S. and simply reassemble them in Holland. A final contract for the fourteen Douglases was not signed until August 15, 1935. In the end, KLM was forced to pay a base price of 145,827 guilders ($88,380) each, which put the final delivery price (including shipping and handling) at 166,000 guilders ($100,600). Fokker's contractual profit on the base price alone amounted to $215,000.[43]

By the time the final contract was signed, relations between Plesman and Fokker had hit rock bottom. In May 1935, Donald Douglas, his wife, Charlotte, and their son, Donald Jr., arrived in Holland for a visit to his European agent and his European launching customer. He instantly hit it off with Plesman, and inevitably, the drawn-out agony of the price negotiations with Fokker came up. When Plesman sketched the state of affairs between him and Fokker, Douglas said he did not believe he had given Fokker the European sales monopoly Fokker claimed to have; rather, he thought, Fokker represented Douglas Aircraft as an independent merchant. Plesman went white with rage. He had Donald Douglas draw up an official statement before a notary to the effect that Fokker held no monopoly rights.[44] That very afternoon, May 23, Plesman told his Board of Control that KLM would never again buy a single Fokker plane and would henceforth become a strictly Douglas operator.

On the other side of the fence, Fokker was just as angry when he learned about Douglas's statement of May 23—after all, he did have his agreement with Douglas Aircraft in black and white. On June 5 Fokker stormed into the Dorchester Hotel in London, where the Douglases were staying. A nasty confrontation between the two men followed. Fokker jabbed his finger at the contract they had signed on January 15 and threatened to take Douglas to a U.S. court and sue him for every penny he owned. Faced with the makings of a big scandal, Douglas—who had also just unwisely accepted an offer from the French constructor Marcel Bloch to buy the licensed manufacturing rights for the DC-2, which had already been sold to Fokker—backed down. Still, Fokker made a claim against Douglas for damages of $50,000 (half of what Fokker had paid Douglas six months before) for dealing with Bloch behind his back.[45]

Relations between Fokker, Douglas, and Plesman thus stood at an all-time low when KLM saw three of its aircraft crash within a week. On July 14, 1935, one of KLM's three F.XXII's crashed during takeoff from Schiphol Airport. Three days later a DC-2 flew into the ground at Bushire, Iran, and three days after that a second DC-2 went down over the Swiss Alps near San Bernardino. There were no survivors in any of these accidents. Certainly, KLM was going through the most disastrous period in its history so far, having already lost its famous DC-2, *Uiver*, on December 20, 1934 (near Rutbah Wells, Iraq), and a Fokker F.XII near Kassel, in Germany, on April 6, 1935. As a result, anxiety about the latest series of crashes ran very high indeed.

Two and a half weeks later, Dutch newspaper reporters questioned Tony Fokker about the disasters. In the interview, which was carried by most of the major newspapers in Holland, Fokker made no secret of his belief that airline procedures were responsible for at least some of the recent crashes. He repeated what he had unsuccessfully tried to bring forward after the Rockne Crash: that airlines, and their macho pilots, were far too eager to take risks and brave the elements to maintain their timetables. He felt that the airline business had become intoxicated with maintaining schedules, especially since the success of the Melbourne Race, and aircraft flew in weather conditions too hazardous to warrant passenger flights.[46]

This time, Plesman was not the only one who was furious. Outraged, KLM's whole Board of Directors met to discuss the interview. Because Fokker had been questioned in the aftermath of the KLM disasters, his statements were regarded as a direct and public attack—an act of vengeance on Fokker's part, inspired by the ongoing conflict over the DC-2 sales, and coming at a time when the airline's public image was most vulnerable. KLM's chairman, August Wurfbain, telephoned Fokker right there and then, demanding an explanation. Emotions ran high. If the board wished, Fokker told Wurfbain, he was perfectly willing to come over and spell out his opinion about the crashes. The conversation went nowhere, and Wurfbain declined the offer.[47] Yet Fokker did not leave it at that. In a nine-page letter to the board of KLM, he took great pains to specify just what he had meant in the interview and why he had spoken out in public: "I consider myself fully entitled to cast my judgment, indeed I consider it no longer appropriate <u>not</u> to do so, so that it cannot be said of me later that I should have spoken out since I am an expert, but remained silent because <u>I AM AN INTERESTED PARTY.</u>"[48]

The issue continued to cloud matters between Fokker and KLM for months, with the parties no longer on speaking terms. Late in October, when Plesman and Guilonard, on their way back from a six-week visit to the United States, found themselves crossing the Atlantic with Fokker on board the *Bremen*, they could not even bring themselves to sit down with him and talk things over.[49] In December Plesman and Fokker were still quarreling, despite various attempts to mediate. Fokker indicated he was willing to bury the hatchet but demanded that Plesman stop using every possible opportunity to spread doubts on the future prospects of the Nederlandsche Vliegtuigenfabriek and to call the quality of its products into question.[50] Not until Christmas Eve did the two sides finally become reconciled.

## GOVERNMENT INVOLVEMENT

As a result of KLM's dramatic change of course, Fokker's Nederlandsche Vliegtuigenfabriek found itself struggling under the financial burden of having developed three expensive aircraft at the same time (the F.XX, F.XXII, and F.XXXVI), none of which held any promise of recouping the investments in them by 1934. Once more the company was on the verge of collapse. The Netherlands government quickly realized the danger. For a neutral country that considered itself a "middle power," a self-creating aviation industry was a strategic asset worth trying to save. On February 3, 1934, a special investigative committee was installed, made up of senior officials from the ministries of Finance, Waterworks, Colonial Affairs, and Defense. It was named after its chairman, Anton van Doorninck, the retired treasury-general of the Ministry of Finance, and its function was to advise the government on the prospects of nationalizing the Fokker company.[51]

Initially, Fokker saw the creation of the committee as a most unwelcome interference in his business affairs. On April 10, 1934, the first meeting between him; his chairman of the Board of Control, Cornelis Vattier Kraane; and the Van Doorninck Committee was conducted in a quarrelsome atmosphere. The officials on the committee reproached Fokker for his long history of failing to cooperate with military authorities in aircraft procurement. Besides, Fokker military aircraft had been falling behind modern advances in design and performance, and were generally delivered way behind schedule. Fokker defended himself vehemently: the departments concerned *never* consulted with his firm in time to incorporate changes in specifications, and thus in the actual performance of his airplanes. How could he, in such situations, anticipate what they wanted? By the time final orders were placed, it was simply too late for the company to heed all desires. A long list of incidents from the past was discussed, and both sides dug in for a fight. In the end, Fokker repeated that government support was not the answer to his company's needs; all he wanted were aircraft orders.[52] A month later his deputy director—Bruno Stephan—and Vattier Kraane made a passionate plea to be given the order for a series of flying boats for the navy in the East Indies to replace aging Dornier Wals, but the Defense and Colonial departments chose to play it safe and ordered the all-metal Dornier Do-24 instead.[53] Nevertheless, the government did not abandon Fokker completely. A contract was signed for a series of thirty C.X light reconnaissance bombers for the Air Corps, which would keep the factory going while the government studied the nationalization issue. As a result, employment at the factory went up again to around 435 people in June 1934.[54] Yet the essential problem—design modernization—remained unresolved. Fokker complained: "The decline in turnover, and thus in revenue, makes it harder every day, and in fact impossible, to find the necessary means to continue the process of development and progress, and make them available above and beyond the general expansion and refurbishing of [the factory's] means of production."[55]

With the work on the F.XXXVI and the F.XXII winding down at the end of the summer, Fokker was already laying off personnel again. In September, fifty people were sent home, and a further 100 to 125 layoffs were considered imminent. Despite claiming to lack interest in government support, on November 2 Fokker filed a request for a subsidy of one million guilders from a special government program aimed at boosting the crisis-stricken Dutch economy. He intended to use the money to build a new factory at Schiphol Airport, equipped with the latest machinery for producing all-metal aircraft.[56] Asked for its advice, the Van Doorninck Committee was skeptical: *if* the government decided to come to Fokker's aid, it would have to exercise tight control over the company. On instructions from the government to negotiate a takeover in one form or another, the Van Doorninck Committee threw out a feeler to Tony Fokker. Worn down by constant fighting in 1934, he indicated that he was willing to discuss the matter, as long as it was understood that he would be kept on as the firm's principal constructor. The Van Doorninck Committee was faced with the problem of maneuvering Fokker into a position where he would have no real control in the company any more. In a meeting on December 12, 1934, Anton van Doorninck himself explained why:

> Fokker must leave his present factory . . . because he is not fit to be director. . . . It is well known that Fokker is not fit to be director. The morale amongst the employees is low. The factory runs better when Fokker is not present. Yet, Fokker has other extraordinary qualities, which we would like to keep. . . . The most likely course would be to keep him on as director, but we have heard from all sides that he is no good as such. Nobody is satisfied with him. There is no order, no regularity in the factory. It suffers from the vices of every company that is badly run. So, if we install Fokker as the director of the new factory, it will run aground. Nevertheless he has good qualities as a constructor and as a businessman.[57]

These were harsh words about someone who had been leading one of the world's most renowned aircraft companies for more than twenty years. Yet Fokker's star had been unmistakably in decline for a long time. The aircraft that carried his name had lost their markets, and his remaining company in Amsterdam was shrinking rapidly, seemingly heading for collapse. The only consolation was the revenue generated by the sale of DC-2's. In 1934, for the first time since 1930, the Nederlandsche Vliegtuigenfabriek was able to close with a profit, however modest: 98,437 guilders.[58]

In a meeting on February 28, 1935, Van Doorninck gently suggested to Fokker that he take up the vaguely defined position of "technical and aeronautical advisor," stressing that it would relieve him of the chores of financial management and leave him free to devote all his energies to aircraft development and sales. To everyone's surprise, Fokker agreed without objection. Whatever they had in mind for the company was fine by him, he told the committee, promising to be "the most cooperative person in the world" as long as he was not forced "to work in the factory from nine to five."[59]

Fokker now seemed certain to lose control over his Dutch company, but he did not really care all that much any more. After five years of near-constant struggle, first in the United States, and then in Holland, he was exhausted, ready to leave the now unrewarding field of aviation behind and enjoy some peace and quiet at last. While there was time—before the Van Doorninck Committee could finalize its proposal for governmental takeover—Fokker instructed his financial manager, F. Elekind, to withdraw 3 million guilders from the company's reserves (then totaling 3.8 million) and transfer it to one of his private accounts.[60] For Fokker, this move served a dual purpose. On the one hand, it safeguarded him against possible losses of the company in the future, when it would no longer be his alone and when it would be difficult to withdraw such funds. On the other hand, reducing the reserves put extra pressure on the Van Doorninck Committee and the Netherlands government to step in quickly.

In Amsterdam, Bruno Stephan was furious about these developments. Though he had guided the company faithfully and stood by Tony Fokker over the years, despite inevitable personal differences, the often careless way Fokker repeatedly treated the company's business as incidental to his private interests had been a source of growing annoyance to him. The latest turn of events gravely endangered the process of modernization and led Stephan to accept an offer from the Turkish government to become its special advisor for the expansion of Turkey's military aviation. In August 1935 he signed a contract in Ankara and left, without remorse, three months later.[61]

Meanwhile, Fokker and the Van Doorninck Committee were moving towards a final agreement on the terms of the government takeover, which included an investment by the state of one million guilders and providing a place to build a modern factory. But much to the dismay of Van Doorninck, Fokker, returning from a restful stay at his chalet in Saint Moritz, suddenly changed his mind again in January 1936 and claimed the directorship of the new company.[62] Once again, Fokker proved himself to be a skillful negotiator: the final agreement stated that he would be allowed to keep the title of director (though without much real influence), with an "advisory" role in aircraft construction and foreign sales, for ten years. He would be paid a modest annual salary of 18,000 guilders for his services, plus the odd figure of three-fifths of 25 percent of the net annual profit.[63]

Only governmental approval was needed now. But when the freshly appointed chief treasurer of the Ministry of Finance read the final report of the committee and its recommendation for government investment in the Nederlandsche Vliegtuigenfabriek, he noted that, in the long process of study and negotiation, one crucial factor had been ignored: "The question is, whether there are sufficient safeguards that the new factory will be able to deal with foreign competition in its engineering capabilities in the long run. . . . Regarding that question, it has to be kept in mind clearly that the cooperation of Mister Fokker as such, does not secure this at all, especially not since the Fokker sun is setting, and besides, one should not be dependent on Fokker alone."[64] Thus warned, the cabinet reevaluated the whole issue carefully, finally deciding on February 22, 1936, that the risks involved were too great. The Fokker company would have to fly on its own wings, or fall.[65] Just when he was ready to step out of the field of aeronautical construction, Tony Fokker was forced back in again. On June 6, 1936, he celebrated his twenty-fifth anniversary as a licensed pilot, still in full control of his company. As a token of appreciation, his staff built a replica of the 1911-model Spider for the occasion. Fokker was delighted with the gesture and relished helping to prepare the machine for flight, even though he prudently left the actual flying to his test pilot, Emil Meinecke. By royal decree, he was also decorated as a knight in the Order of the Netherlands Lion.[66]

# 9

## ROCKING THE BOAT

If there is one thing that can be said for Tony Fokker's professional activities, it is that he had a keen sense for business, instantly recognizing opportunities to make money. When, late in 1933, the outlook for the Dutch Fokker company suddenly paled, he took great care to seal off his traditional export market before the rapidly expanding American civil aircraft manufacturers readjusted their sights on the worldwide market. If KLM, then probably Europe's most cost-conscious airline, paid instant attention to the DC-2, other leading carriers might soon follow. Early in January 1934, he arrived in Los Angeles for discussions with Donald Douglas. Aware of the impact that the latest American-built airliners might have on Europe, Fokker made sure to close a sales and license-manufacturing agreement. In itself, this was unusual: one established manufacturer representing the interests of a competitor newly entering his own "home market."

Yet Fokker did not stop there. In mid-January, while finalizing the DC-2 contract with Donald Douglas, the latter introduced Tony Fokker to Elliott Roosevelt, son of President Franklin D. Roosevelt and a noted aviation enthusiast with his own pilot's license. Elliott Roosevelt and his business partners, George W. Stratton and Franklin K. Lane, had arrived in California a little earlier than Fokker and had bought the Russian sales rights for the DC-2, anticipating what Roosevelt described as an extremely well-funded Soviet aeronautical purchasing commission, due to arrive in the U.S. in February. The Soviets were reportedly seeking to acquire fifty or more modern airliners.

Fokker's attention was aroused: was there a way he could participate in the lucrative deal that Roosevelt had told him about? He needed a competitive airplane to do so. Always quick to see an opportunity for business, Fokker walked into the office of Robert E. Gross, the top man of the Lockheed Aircraft Corporation in nearby Burbank, the day after his agreement with Douglas. Aided by a $200,000 loan from the Reconstruction Finance Corporation, Lockheed was then just finishing the prototype of its own high-tech approach to civil aircraft engineering, the L-10 Electra. The Lockheed plane—like the DC-2, designed to compete with the Boeing 247—was the smallest and fastest of the three, seating eight to ten passengers. Gross hoped that its size would make the Electra attractive to airlines and executive business use.[1] Making shrewd use of the fact that the Electra had not yet flown and was thus an unproven design, Fokker negotiated for a European sales agency with an exclusive option to buy a manufacturing license later. Though cautious at first, Gross was soon won over to the idea of having the well-established Fokker organization backing his latest venture, and within a week Fokker and Gross reached a tentative agreement. Fokker ordered one Electra and paid a cash deposit for it. He was granted exclusive sales rights to the Electra for two years in Europe and Great Britain and its dominions, except Canada. But what he had really come for were the rights to sell the Electra in the USSR. This necessitated some difficult legal fine-tuning. Gross was well aware of the financial risks involved in aircraft sales to the USSR, not the most trustworthy of countries when it came to making payments. Lockheed also had various commitments for its earlier Orion, Vega, and Altair types, and their sales to the Soviet Union had to be considered. But after a week of careful study, Gross let Fokker have his way because the deal involved welcome extra cash flow for Lockheed at no risk to the company.[2] Early in February, the agreement was finalized. Because there were, at that point, no discussions of European orders for the Electra, Fokker obtained the rights at the bargain price of $16,000.[3] With Boeing's production tied up by United, he was now well poised to remain at the center of Europe's aviation business, regardless of the fortunes of the Nederlandsche Vliegtuigenfabriek.

After the business with Lockheed had been sealed, and while recuperating in the southern California sun from an appendectomy, Fokker pondered the possible sales to the Soviet Union. Reasoning that Roosevelt might well be able to use his position as the president's son to gain the best possible introduction to the Soviet commission, he had his legal council and personal business manager, Carter Tiffany, approach Elliott Roosevelt in New York with a proposal to join forces. If Roosevelt could also draw the Soviets' attention to the Electra, he would thus have two irons in the fire and be all the more likely to strike a deal—one that could be very beneficial to Fokker's stake in the Electra.

On February 28, 1934, Fokker and the Roosevelt team agreed that if the Soviets indeed bought fifty Electras, they would split the profits on a fifty-fifty basis: $500,000 each. Before a U.S. Senate investigative committee, Fokker later confided that he had thought such a profit excessively high, but that Roosevelt assured him he could swing the deal because of his influence with the Soviet purchasing commission and the Import and Export Bank. Fokker paid Roosevelt an advance of $5,000 in cash and a check for $6,666 for Roosevelt to travel to the Soviet Union, where he would make full use of his father's name and position to gain access to the highest levels in the Kremlin. But the president felt this was taking things a little too far, and he put a sudden stop to the scheme.[4]

Even after the arrangement was short-circuited in Washington, Fokker remained optimistic that a number of Electras could still be sold to the Soviet Union. Early in March he arrived in Amsterdam, still contemplating the Soviet possibilities. During the next month, he traveled around Europe, discussing the sale of a sublicense for the DC-2 with the Swiss branch of the German Dornier company, and he extended a similar offer to the Svenska Aero AB in Lidingö, Sweden. In Madrid, he proposed to supply a troop transport version of the DC-2 to the Spanish government.[5] None of these initiatives led to contracts, but Fokker did sell six Electras in France and Spain, for which Carter Tiffany deposited $58,500 in cash with Lockheed.[6] The Lockheed rights seemed to be paying off.

On May 16, 1934, Fokker left Holland for a trip to Moscow—a welcome opportunity to get away from the high-pitched confrontation with KLM over the modifications necessary for the U.S.-built DC-2 to meet Dutch requirements for a certificate of airworthiness. The Soviet trip would allow things to cool down for a while. He hoped to secure a contract for the Lockheed Electras in Moscow. Arriving in Moscow, Fokker saw a country very much changed from the Russia he had visited in 1911. On the surface, these changes appeared to be for the better. In a Dutch newspaper interview, he commented in an upbeat spirit, "I have flown over a large part of Russia and have seen the signs everywhere of a new country on the rise, and the powerful will that is present to shake the old system and become a big and mighty country."[7]

Just how powerful that "will that is present" had become, Fokker, like other western visitors, had little idea. In 1934 the Soviet Union was a country suffering from the effects of "Stalin's Revolution": a forced transition of the economy, and of society, under the first of Joseph Stalin's Five-Year Plans (1929–34).[8] Stalin's program, which uprooted the whole society during that time, consisted of two main streams: a ruthless collectivization of the agricultural sector, and a rapid expansion of heavy industry. To achieve these goals, the Soviet populace suffered severe material hardships, on top of which Stalin cloaked the country in an atmosphere of extreme insecurity and repression, aided by a constant fear of the powers of the state. Since 1929 Stalin had succeeded in concentrating these powers almost exclusively in his person; they were maintained through the Communist Party and the secret police, the NKVD. Policy and production targets changed repeatedly, and Stalin's changes were achieved only at the cost of millions of lives. Nevertheless, the Soviet media boasted of a substantial increase in industrial production and the initiation of a number of large-scale public works projects, symbolic of the new socialist era. The establishment in 1932 of the Soviet airline Aeroflot, which grew to be the largest airline in the world, was stimulated by the first Five-Year Plan. In January 1934 Stalin proclaimed success in "the struggle to overcome Russian backwardness."[9] Foreign visitors, shown only the material manifestations of these attainments, were generally impressed, and Tony Fokker was no exception. What made Fokker's visit interesting was that it provided the only occasion when he publicly took issue on international politics. Failing to note the effects of Stalin's repressive regime except in the manifest shortage of goods, Fokker spoke out in favor of diplomatic recognition of the USSR by the Netherlands in the interest of closer commercial ties between the two countries. He told reporters that he expected benefits for Dutch exports from such a move,[10] anticipating that large markets would soon open up for "household commodities like furniture, lamps, paper, writing materials, radios, medicines, sanitary equipment, tinned foods, textiles, metal objects, photo cameras, electrical appliances, car parts, shoes (though not army boots), etc., in short, goods for daily life that make our lives in western Europe more agreeable, but of which one sees very little evidence in the Soviet Union."[11]

Despite prior assurances from Elliott Roosevelt that a Soviet order for the Electra was imminent, Fokker returned from his eleven-day trip empty-handed. Having arrived in Moscow without the necessary contacts lined up, he discovered to his frustration that he stood precious little chance of getting through to the decision-making levels of Stalin's bureaucracy. He arrived back in Holland with a temper on May 28 and wrote an angry letter to Donald Douglas about Douglas's elite "sales representatives" (the Roosevelt team), whom Fokker described as "chisellers."[12] Two weeks later, on the basis of his own company's long-established relations with the Polish airline LOT, Fokker was rather more successful, selling two Electras in Poland and closing a sublicense agreement there.[13]

The difficulty of selling airplanes was not Fokker's only frustration. With prospects for the Nederlandsche Vliegtuigenfabriek deteriorating rapidly as KLM plotted a new course in aircraft development, 1934 was turning into a nightmare year. He knew he would have to look for other ways to maintain a position of prominence in international aviation.

In October 1934, an unexpected opportunity arose when Fokker was approached by a small British firm, Airspeed Ltd. Airspeed had literally just outgrown the garage in York, where its director/engineer (and future best-selling novelist), Nevil Shute Norway, and his partner, Hessell Tiltman, started out in 1930 and was now seeking an association with an established name in aviation to expand its business.[14] After a successful first public stock issue, and in the wake of the success of the *Uiver* in the Melbourne Race, Airspeed decided to approach Fokker. After all, he carried the license-manufacturing rights to the DC-2. Fokker, on his part, saw a welcome opportunity to make some money on his old patents and designs before they lost their value. In a series of discussions—conducted, in typical Fokker style, at odd hours, in empty restaurants in London, Newcastle, and Amsterdam, and in Fokker's Oberalpina chalet in Saint Moritz—Fokker, Norway, and his new codirector, George Wigham Richardson, hammered out an agreement on January 25, 1935. At a cost of £20,000, Airspeed acquired the right to license-manufacture no fewer than fifteen types of Fokker aircraft for sale in the United Kingdom; all territories of the British Empire, including Canada and other dominions; and the League of Nations Mandates, governed by Britain. It sounded impressive but yielded Airspeed preciously little in actual practice. Norway and Richardson were really after the sublicense for the manufacture and sale of the DC-2, but Fokker limited this right to the United

Kingdom, the Channel Islands, the Isle of Man, and the Irish Free State.[15] Norway's hope that Fokker could somehow be instrumental in expanding Airspeed proved futile. Despite its hopes, the company was unable to overcome British import barriers and never sold a single DC-2. Fokker's aspiration for the deal to pierce the trade barriers the British Empire had erected in 1932 as part of its effort to combat the world economic crisis was also dashed. Consequently, the Fokker-Airspeed collaboration did not amount to much. Fokker, who had himself appointed "technical and aeronautical advisor" to Airspeed, admitted that the work load imposed on him was "minimal."[16] The Airspeed contract was the last time Fokker was able to sell the rights to license-production of his products, and the last time he was sought out as an aeronautical consultant. The experience must have been gratifying for him amidst the difficulties he was experiencing in 1934 and 1935. In 1936 the agreement was quietly terminated when Airspeed contracted Air Ministry work under the British rearmament program and was forbidden to admit foreigners on its premises by the Official Secrets Act.[17]

The Airspeed deal sounded a few upbeat notes in a sad melody, but the *marche funèbre* of the Fokker company was far from over. At KLM, Plesman made it clear beyond the shadow of a doubt that he stood by his resolve never to buy Fokker aircraft. For Fokker, such a decision was unacceptable. Having invested comparatively large amounts in the development of the F.XXXVI and F.XXII, so that the four-engined airliners had practically sealed the fate of the company, it was impossible simply to abandon the designs. He therefore proposed to develop an improved version of the F.XXXVI, the F.37 (Fokker henceforth adopted more easily digestible arabic numerals) as a future replacement for the DC-2's on KLM's East Indies route. The F.37 was to have a retractable landing gear and incorporate substantial aerodynamic improvements over the F.XXXVI, thus closing the gap in design efficiency between it and the DC-2.[18]

Plesman refused even to discuss the matter seriously, and Fokker's prospects declined further as a result of the American trip Plesman and Guilonard made in September and October 1935. Special guests at Douglas Aircraft, Plesman and Guilonard came back thoroughly acquainted with Donald Douglas's new commercial design, the DC-3, and his preliminary ideas for a very large four-engined airliner, equipped with a pressurized cabin, and capable of operating at high altitudes. Fokker could offer nothing comparable, and in a meeting on November 11, 1935, an enraged Fokker was told that KLM was looking at Douglas's four-engined airliner for the airline's long-term planning. (Plesman was enthusiastic enough about the aircraft to contribute 350,000 guilders toward the DC-4's research and development.)[19]

Fokker did not give up easily. Studying the preliminary data on the Douglas project, he came up with Design 127 (later referred to as F.56) not three weeks later: a four-engined airplane capable of carrying fifty-six passengers in a fuselage divided into an upper and lower deck. Fokker promised he could build such a plane much earlier than the Douglas design would become available if KLM accepted its mixed wood and metal construction. In the meantime, he proposed that KLM order his F.37 for continued capacity expansion on the Amsterdam-Batavia route, rather than the DC-3, which Plesman was considering.

But the days of designing new aircraft in a matter of weeks had passed. To promote a really competitive airplane, Fokker would have needed far more time than he allowed his design staff, and the "finalized" plans for the F.56 that were sent to KLM on February 6, 1936, were quickly rejected as falling far short of what Douglas offered. In the following years, the frustration of all those at the Nederlandsche Vliegtuigenfabriek who dealt with KLM mounted continuously. None of the designs they developed and proposed to the airline (designs 132, 133, 136, 138, 139, 140, 141, 144, 145, 147, 158, 159, 173, and 191) met with anything but, at best, polite rejection.[20]

Besides, the operational experience KLM was accumulating with the DC-2 further strengthened Plesman's resolve to stick with Douglas. In May 1936, three months after the Netherlands cabinet had finally decided not to invest in the Fokker company, KLM received permission from the government to buy one DC-3, the prototype of which had made its first flight five months before, on December 17, 1935. At that time, there had been a slight prospect that Fokker would license-manufacture this new type. But when it became clear that such was not the case, KLM was given permission to order twenty-two more DC-3's.[21] Like the DC-2's, they had to be acquired through Fokker, but fortunately for both parties, Fokker, spending less and less time in Holland, decided to remain aloof from the negotiations over their sale.[22]

Instead, he focused his attention on a different aspect of aeronautical development: airport planning. A year and a half before, the Amsterdam City Council, undecided about how to expand ground facilities at its municipal airport, Schiphol, had asked Fokker about his views on future requirements of airports. In May 1936 he came up with a futuristic, detailed plan for a large airport with a circular terminal building on a "traffic isle" on the airfield, the center of a tangential system of (grass) runways. Fokker's detailed design allowed maximum efficiency in aircraft handling and capacity for loading and unloading more than two million passengers per year. In his report he stressed the need to adopt a long-term policy in airport construction and take into account the growth of air transport and the future would have in the sense of demand for purpose-developed ground facilities. He concluded: "Airport construction . . . demands a wide view of the future and a broad-based approach to the issues at stake. If anywhere, it is here that frugality deceives wisdom. Presently refusing to clearly face the problems that must be dealt with, and resolve them for a number of years, will avenge itself by necessitating enormous new investments later."[23] Fokker's plans for a new airport were initially shelved because of a lack of funds, but they constituted the start of the planning process for a new airport, which came to fruition thirty-three years later, in 1967.

## BIDING TIME

Tony Fokker never enjoyed the satisfaction of being praised for his far-reaching ideas. To many onlookers, he seemed to become more and more reclusive. The events of 1929 left deep marks on his life. With Violet's suicide, the bottom had dropped out of his personal affairs. Coming to terms with her death was very difficult, and his distress was all the more intense as it became evident that his position in the restructured Fokker Corporation of America was also far from secure. Life was not treating Fokker too kindly, and in the early 1930s, even his health, which had never posed a serious problem, started playing tricks on him. He was beginning to pay the price for his many years of flying in open cockpits, unhealthy eating habits, and the physical demands of his irregular lifestyle. His sinus condition was gradually getting worse, and his lungs were starting to give him trouble too with repeated attacks of pleurisy, a serious disease that restricts breathing because of excessive fluid in the lungs.

The *pilote par excellence* was pushed into the role of passenger when he flew. For Fokker to take the controls himself and demonstrate his extraordinary flying skills, the occasion needed to be very special indeed. On July 2, 1930, Charles Kingsford-Smith, whom Fokker had entertained with his crew after their successful westbound flight across the Atlantic, took off in his *Southern Cross* from Roosevelt Field in New York, bound for California to complete his round-the-world flight. Fokker, already in a buoyant mood, had just spent a gratifying week relaxing in the company of Kingsford-Smith's people and a bunch of other friendly aviators. But what made the occasion really special was the arrival of a small monoplane in the middle of the festivities, on the morning of Kingsford-Smith's departure, from which his old-time Russian sweetheart, Ljuba Galantschikova, emerged. She now called herself Ljuba Phillips, keeping the name of her American husband, whom she had divorced. Fokker was on cloud nine in the company of the ever-vivacious Ljuba. With Tony at the controls, they took off together in her airplane to indulge in a feast of Johannisthal nostalgia. Fokker pulled out all the stops to impress his former girlfriend, offering her, and the bewildered spectators at Roosevelt Field, a hair-raising demonstration of low-altitude aerobatics. But while he and Ljuba were having a good time, the Department of Commerce supervisor at Roosevelt Field was not. Several times he ran toward the airplane when it appeared to be landing, only to find out that Fokker was performing "touch and go" maneuvers. When the airplane finally came to a full stop, Fokker emerged, still in high spirits, only to be told to turn over his pilot's license immediately, having broken just about every regulation in the book. When it became apparent that he no longer had a valid license—it had lapsed long ago, and it had never occurred to him to renew it—Fokker, like a bad boy, was summoned to report to the supervisor's office and fined $500 for piloting a plane without a license, despite his protests. When Tony offered to make amends, he was given a "letter of authority to fly under student regulations" after a brief medical examination, pending a formal exam.[24] His license was later restored to him without an examination—a tribute to his fame.

The incident exemplifies Fokker's careless attitude to the formalities of life. When his friend Bernt Balchen asked him to testify before the Bergen County Naturalization Court in Hackensack, New Jersey, to support Balchen's application for citizenship papers, it dawned on Fokker that he had completely forgotten about his own case, pending since 1926. Unable to testify himself as a noncitizen, he nevertheless accompanied Balchen to the courthouse, and they concluded their procedures together on June 26, 1931, with Carter and Anne Tiffany as their witnesses. Three months later, Fokker's U.S. citizenship papers finally arrived.[25]

By that time his role as an American aircraft constructor was already over. On July 10, 1931, having been given a golden handshake as he left General Aviation, Fokker and the Tiffanys went on board Tony's motor yacht, the *Honey Moon,* for a long cruise in Long Island Sound. The ship was one of the few continuities in Tony Fokker's life. He was extraordinarily attached to it, and had it transported to the U.S. after his move in September 1925. Though the *Honey Moon* was usually moored in the marina of Montauk, Long Island, Fokker kept the ship on the Netherlands register and proudly flew the Dutch flag from its stern—a rich man's prerogative. Until he tired of flying and moved the ship to Englewood Basin marina, he covered the 130 miles from Alpine-on-Hudson to Montauk in his amphibious flying boat. The *Honey Moon* embodied his finest memories. He had spent his happiest times aboard the yacht, first with Tetta, and later with Violet. Even after Violet's death, when the name became a sarcastic allusion to his misunderstanding of women and his mistake of confusing temporary relaxation with deeper marital happiness, he found himself unable to let go of the ship and the memories it stood for. Instead, he renamed it *Helga.* He spent more and more time aboard as the years went by, fishing, pondering the many problems of his professional life that chased him out to sea, and gazing at the stars on long, sleepless nights. For more excitement, he kept a small "utility speedboat," *Yo Ho Ho,* from the bow of which he liked to hunt sharks and dolphins with harpoons. Boating continued to be his favorite escape from the silence of his somewhat gloomy house in Alpine-on-Hudson, with its dark, massive Dutch and German furniture and old paintings by Dutch Masters that he had surrounded himself with over the years, and from the repetitive pressures of business.[26]

While on one such sailing trip, recuperating from the clashes with KLM and the stress of dealing with the Van Doorninck Committee, Fokker was subpoenaed to appear before the U.S. Senate Special Committee Investigating the Munitions Industry. Headed by Sen. Gerald P. Nye, a Republican from North Dakota, the committee conducted hearings on arms and munitions exports from the U.S. Hearings ran between April 1934 and July 1936 and investigated the involvement of a great number of U.S. firms in armaments exports from the early 1900s to 1934. Because the committee members had heard various rumors about contacts between Fokker, Elliott Roosevelt, and the Soviets, the Nye Committee, as it was commonly known, wished to question Fokker. The involvement of the president's son introduced distinctly political nuances to Nye's interest in the matter, too. When the process server called at

Fokker's home on August 30, 1935, he found it empty, having missed Fokker by a few hours. Alarmed, suspecting that the committee was on to something big and that Fokker might well be on the run, Senator Nye requested a full-scale police search. Posts on the Canadian and Mexican borders were alerted, and Nye informed the State Department.[27]

Aboard the *Helga,* off Montauk Point, Fokker was oblivious to all this consternation. He had recently returned to the U.S. in the company of a Dutch friend, Wim van Neijenhoff, and was spending his days shark-hunting and trying to film whales, which were especially numerous in the waters around Long Island that summer. The judiciary took several days to locate him, and by that time, the press had gotten wind of the story. Fokker was the last to find out he was a wanted man: he read it in a newspaper on September 4, having returned to Montauk from a shopping trip in nearby East Hampton. Around 7:30 P.M. that day, he stepped out of his car at the Montauk quayside and was bewildered to be apprehended by process server Stanley Ferguson, specially arrived by train from New York. Because it was clear that Fokker had no intention of fleeing the country, the matter of the subpoena was quickly resolved, and it was agreed that he would appear before two members of the Nye Committee after his holiday, on September 12.[28] Fokker duly showed up in the company of Carter Tiffany, his legal council, armed with a collection of documents showing that no arms sales had been intended in the contacts with Moscow. Nye had caused a storm in a teacup, something he seemed to have a special gift for. In October 1936 the senator felt obliged to publish the full text of the Fokker proceedings after the aeronautical press accused him of withholding information from the public on the affair. This decision gave rise to a short-lived partisan quibble between Democratic and Republican committee members.[29] In the published, 20,000-page report on the committee's investigations, there was not a single word about the Fokker episode. Fokker himself, however, did not avoid mentioning Elliott Roosevelt in an interview with the *Los Angeles Times* about the Soviet affair, and as a result, Roosevelt threatened to sue for slander.[30]

Such upheavals did not diminish Fokker's activities as a middleman in aircraft sales, though his role was often curiously aloof. In July 1936 he appointed Richard M. Mock, in Hollywood, California, to represent him in regular contacts with Douglas. Mock's job was to maintain a close working relationship with the manufacturer and monitor the latest technological developments, though, in typical Fokker fashion, he doubled as a kind of personal secretary. On Fokker's instructions, Mock closely watched the construction and testing of the first DC-3 conversion of the DST (Douglas Sleeper Transport) and the building and testing of a Douglas flying boat. He also bid for the sales rights to the Douglas DB-1 bomber in Europe and the Netherlands East Indies. Mock's job was not an easy one. His authority was strictly limited, and he frequently needed specific instructions, but Fokker was usually difficult to find and hated being chased by business mail.[31] To overcome this communications problem, Mock sent telegrams to Carter Tiffany at Tiffany's office in New York, asking him to telephone Fokker at home in Alpine-on-Hudson and relay messages about various developments. He also sent telegrams to a telegraph office near Fokker with instructions for the operator to telephone Tony at regular intervals and convey his message. In the same roundabout way, Fokker occasionally wired Mock back to let him know when he would call with instructions. Typically, he phoned after 11:30 P.M. or around 8:30 A.M. New York time, keeping Mock tied to his office telephone around the clock.[32] In July 1936 Mock had to track him down at the Montauk Yacht Club for instructions on how to go about reserving assembly-line numbers for an order of seven DC-3's for KLM—a problem that arose because Douglas was already committed to producing the DC-3 for several U.S. domestic carriers and thus had to try and squeeze the Fokker/KLM order in. Like the DC-2's before them, the DC-3's were transported to Europe almost completely assembled. Though this method involved more risk of damage, Fokker preferred to avoid paying for disassembly at Douglas (after the completed aircraft had been test-flown and checked) and reassembly in Holland.[33]

Mock did as much of the work for the European Fokker-Douglas sales as anyone, but he was often kept in the dark about Fokker's motives or strategy. Always keeping his ear close to the ground, Fokker had heard rumors that a delegation from the Polish airline LOT was expected to visit the U.S., and was looking for an inexpensive replacement for their aging Fokker F.VII's. Fokker imagined that an updated version of the Consolidated Fleetster might be just what they wanted and immediately began bargaining with Consolidated for European sales rights in September 1937. Mock only heard about it when he accidentally bumped into KLM's technical manager, Henk Veenendaal, who was posted at Douglas at that time.[34] Mock was instructed to accompany the Poles as they visited the various manufacturers in California (Douglas, Lockheed, and Consolidated) and make sure that any quotations of prices and delivery terms were the same as those of the Nederlandsche Vliegtuigenfabriek

Fokker.[35] The Polish delegation ordered five Lockheeds for LOT and were promised delivery in February 1938.

With Mock holding the fort in California, Fokker's activities as an aircraft salesman took up only limited amounts of his time. That time—or rather the control over it—became more precious to him every year. An avid photographer, Fokker spent much time traveling for pleasure across the Atlantic on the ship of his latest preference, the *Queen Mary,* and within the United States. In September 1936 he visited the Grand Canyon National Park, where he stayed at the El Tovar Hotel on the South Rim and hiked on one of the trails to find the best perspectives for landscape photography. He was still fit enough for hiking, but his age showed when he discovered he was seriously handicapped without his glasses, which he had left at the hotel. Later, on Fokker's repeated instructions, Mock went to great lengths to discover their whereabouts and have them sent back to him in Alpine-on-Hudson.[36] Fokker had worn glasses since an eye operation at the Eye and Ear Hospital of Los Angeles a year before,[37] a minor indication that his health in general was beginning to suffer. After a trip to San Diego in September 1937, he went home with some kind of infection, feeling so ill that he had to telegraph Richard Mock to confer with one of his trusted Los Angeles specialists, a Doctor Jesberg, who advised him to see one of his colleagues in New York straight away.[38]

Fokker continued to spend the winters in Saint Moritz, but sailing occupied a major part of his American sojourns. Indeed, for several years, Fokker had been working on what he came to consider one of the main achievements in his life: his own design for a new ship to replace the *Helga.* In this 112-foot yacht of very unconventional appearance, he attempted to blend radical streamlining and the power of a speedboat with the comfort of a conventional yacht. True to the materials he knew best in aircraft construction, the yacht's hull and superstructure had a plywood outer shell. The ship was powered by three engines, one conventional maritime diesel engine and two specially adapted 600 hp Wright Typhoon aero engines for high-performance operation. To accommodate Fokker's sharking hobby, the ship was designed to carry a small, high-speed fishing craft and a power tender inside the ship's hull, which could be put overboard by retractable davits. In the autumn of 1937 the design, which he had worked on intermittently for five years, was nearing completion. Typically, a full-size mock-up was constructed of cardboard before the blueprints were completed. At the National Motor Boat Show in New York, on January 13, 1938, Consolidated Shipbuilding, on the Harlem River, proudly announced that it had secured the order to build it.[39]

For Fokker, the construction of the ship was the culmination of a life-long ambition, and he practically lived at the shipyard.[40] The ship focused his attention on maritime activities more than ever. When he finished the design, he decided to move so he could have his own private mooring facility and immediate access to the ship at all times. In July 1937 he found precisely the house he was looking for in Upper Nyack, twelve miles upstream from Alpine-on-Hudson. Here the cliffs above the river gave way to a steep, soil-covered slope, where a picturesque village sprawled along the river's edge—a fitting environment for New York businessmen to escape from the rigors of city life. Fokker bought a roomy, red-brick house in Nyack on North Broadway, called Undercliff Manor—an apt name, for it was located at the bottom of a 300-foot-high cliff. The house suited Fokker perfectly. Unobtrusive when seen from the roadside, it had large windows at the back and a beautiful view of the Hudson River that was his pride and joy. Fokker changed the large central hallway of the house, breaking out the back wall and replacing it with a columned, glass-enclosed portico. Seen from there, the panorama of his terraced garden and the Hudson left first-time visitors in awe.[41]

On June 20, 1938, the new ship was finally ready for launch. It had cost Fokker $200,000—more than he ever spent to design any of his airplanes. At 3 p.m. he addressed the numerous guests at the Harlem River quayside in a boisterous mood, beaming with pride:

> I know that many yachtsmen are skeptical of this ship. That is the way I want it. I wish to prove to these skeptics that my ship, by introducing new principles, will revolutionize, and give new impetus to the shipbuilding industry. I do not think my ship will last for ever. I hope it will be obsolete within two years. That is the way we build airplanes. No sooner do we complete them than they are obsolete. That is good. That is progress. There are too many yachts which outlive their owners.[42]

At seven minutes past three, Fokker spoke on the telephone to his mother in Holland, and from across the Atlantic, she christened the ship *Q.E.D.* (*quod erat demonstrandum*: thus it has been proven), another sign of Fokker's deliberate defiance of accepted shipping designs. Thereupon he smashed a bottle of *water* that he had brought from Holland on the ship's bow, a fitting gesture for a steadfast teetotaler. On board the *Q.E.D.*, a bell sounded as the ship slid into the water . . . to an immediate and embarrassing halt. Neither Fokker nor Consolidated Shipbuilding had reckoned with the tide. To the amusement of the press, the *Q.E.D.* remained immovably stuck in the muddy bottom of the Harlem River. A nearby tug lost three cables trying to tow the ship free. Fokker waited anxiously until the next high tide, at 3:30 in the morning, before his precious yacht finally floated.[43]

Immensely proud of his unusual vessel, which attracted much attention in maritime circles, Fokker enjoyed the *Q.E.D.* enormously, even though he had to call in a special team of engineers from Douglas to solve the problem of excessive vibration from the ship's three engines. Yet his words about too many yachts outliving their owners would turn out to be distressingly prophetic. In October 1939 he lent the ship to a couple of friends for their honeymoon. In the middle of the Hudson River, near Yonkers, the *Q.E.D.*'s electrical installation short-circuited. The ship immediately caught fire. All twelve people on board jumped into the river, but only eleven managed to reach the shore. The ship itself was destroyed.[44]

The loss of the fruition of a lifelong ambition in shipbuilding was felt deeply by Fokker; yet having walked the towpath of yachting design and found it gratifying, he set about drafting plans for another unorthodox ship with great energy. Shortly after the destruction of the *Q.E.D.* he received a letter from Starling Burgess, a former aircraft constructor, long since turned naval architect. Burgess, who had been most impressed with the design of the *Q.E.D.*, offered his services in case Fokker might be contemplating another radical incursion into shipping design. Four days later, Fokker telephoned him, and a meeting was arranged in Burgess's office. The two men had several things in common. Both had been involved in the early days of aircraft construction, and they shared an enthusiasm for boating. Fokker explained that he was looking for a ship that would embody the ideal proportions of engine power and sail to be used according to weather conditions and his preference of the moment. Fokker visualized a ship with hollow steel masts from which sails could be unfurled at the push of a button. Within a month, Burgess produced three draft designs for a three-masted ship, and near the end of November 1939 the two men were rapidly moving toward some sort of partnership in nautical design.[45]

## CLOUDS WITH SILVER LININGS

After moving to Nyack, Fokker intended to spend much of his time relaxing on the water, but this did not mean that his professional life had dissolved. The decision of the Netherlands government not to take over the Nederlandsche Vliegtuigenfabriek had left him still in charge of the company. It was not a position he particularly enjoyed any longer. The years of endless conflicts with Plesman and KLM, the limited size of Dutch military orders, and the disappointment over the government's decision to pull out at the last minute had embittered him. Refusing to recognize how his decline as an aeronautical constructor had resulted from his financial policies and his own limitations as an engineer, he felt wronged that his contribution to Dutch aviation was recognized only in words, and in the past tense, at that. As a consequence, his personal involvement in running his Dutch company diminished.[46]

Fokker's withdrawal worked out to be a blessing, for it freed his new deputy director, Jacob van Tijen, who had succeeded Stephan in November 1935; designer Erwin Schatzki, a refugee from Nazi Germany who had been hired in March 1934; and chief engineer Marius Beeling to pursue a course of gradual modernization. Van Tijen had started his unusual career as an export manager in the cocoa business. In the early 1930s, his accomplishments in long-distance private flying had brought him an appointment as the chief executive of a Dutch shipping conglomerate studying the possibilities of airship operations. Van Tijen's efforts to bring the Fokker company back into the European aeronautical picture were aided by the tide of international politics, which led the Fokker company to concentrate yet again on developing military aircraft.

Fokker condoned this shift. Though he preferred to steer clear of politics as best he could throughout his life, he had a clear sense of the encroaching dangers of war in Europe. In April 1935, in a conversation with Dutch business tycoon Ernst Heldring, he painted a vivid picture of German preparations for war, particularly in the field of aviation.[47] Three years later he was convinced that war was inevitable and the question of when it would break out depended only on Hitler, "who is doing what he pleases." Echoing the conviction of many people watching the advances in military aviation, he was certain that the next war would be decided in the air, with large-scale attacks by fast, high-flying bombers against civilian and industrial targets. Such air raids, Fokker believed, would prove unstoppable: the bomber would always get through.[48] To be ready for all eventualities, Fokker transferred $2.85 million of his private capital from Holland to the United States in 1938 and 1939, plus $1 million of company reserves.[49]

Despite such transfers, the Nederlandsche Vliegtuigenfabriek was climbing out of recession thanks to the rearmament programs of various European governments. With the major contestants for power in Europe banning arms exports to keep from losing momentum in their own defense preparations, the Fokker factory once more became the focal point of military aircraft procurement commissions. In quick succession, the factory turned out new types in an attempt to cater to all needs. In 1936 a new single-engined fighter appeared, the D.XXI, followed later that year by a remarkably unorthodox, twin-engined, multirole fighter-bomber, the G.1, which first flew on March 16, 1937. A new twin-engined medium bomber, the T.V, which doubled as an "air cruiser," followed only months later. On the insistence of Fokker's Board of Control that capital expansion was now absolutely vital because European export markets for military aircraft were opening up again, Fokker finally agreed to change the structure of the company to allow a fresh influx of capital.

After the company's statutes were changed on May 13, 1937, the stock of the restructured company, now officially called Nederlandsche Vliegtuigenfabriek Fokker, was introduced on the Amsterdam Stock Exchange on May 26. The issue was successful, and the paid-up capital of the company increased from 500,000 to 2,500,000 guilders. This infusion enabled a 52-percent expansion of factory floor space and the first steps toward catching up in the field of all-metal aircraft construction.[50] Employment was picking up again, too, as the factory made use of its increased floor space and mixed construction techniques to speed up production by taking on new woodworkers. By August 1937 Fokker was employing 750 people, a number that rose with increased demand to 1,200 in 1938.[51] An agreement for the license-manufacture of ten D.XXI fighters in Denmark and thirty-eight D.XXI's in Finland had been drawn up a month earlier, while negotiations for future license-production of the same aircraft for the Spanish (nationalist) government were cut short only by the more immediate demands caused by the outbreak of the Spanish Civil War.[52] As a result of these developments, the company was reporting profits again: 1,139,593 guilders in 1937, rising slightly in 1938 and 1939 to 1,299,088 and 1,424,207 guilders, respectively.[53]

Tony Fokker's personal involvement in this rapid, rearmament-driven recovery of his company was only marginal. His influence, exercised mainly during his occasional visits to Holland en route to and from his Swiss winter chalet were, if anything, disruptive. He seemed unable to reconcile himself to the idea of having been dumped by Plesman and KLM. Despite the demands of the military projects now being pursued under Van Tijen's management—a new twin-engined fighter project (the D.XXIII) and an all-metal bomber for the air arm of the Netherlands East Indies Army (the T.IX)—he kept pressing for new civil designs for short-haul and intercontinental airliners, none of which KLM was willing to buy. On February 11, 1939, he addressed a powerful, passionate appeal to the Dutch prime minister, Hendrik Colijn, to force KLM to order Fokker airliners in the long-term interest of the company: "Now the moment has arrived, in which the *position of the Fokker factory on the future world market* can be *influenced decisively*, which should lead to a *recapturing* of its *prominent, independent position* amongst [aircraft] manufacturers, or [witness] a *lasting debasement* to an agency, dependent on foreign ventures."[54]

Fokker reiterated his case several times that month. Senior civil servants in the Netherlands Aeronautical Service pleaded against the request, unequivocally stating it was his responsibility as a manufacturer to spearhead aeronautical developments instead of chasing them, and that it was in the best interest of KLM (and thus of the Dutch taxpayer) to allow the airline to buy the most economical planes available on the international market.[55] But Fokker's direct appeal, coming as it did at a time when the Fokker company was playing a vital role in Dutch defense preparations, could not be left unheeded. Early in March the minister of waterworks ordered Plesman to discuss the various aircraft types he had to offer with Fokker.[56] After several preparatory meetings between mediators from both sides, the two adversaries, accompanied by their legal advisors, finally sat down on March 13, 1939.

The meeting had been arranged to take place on neutral ground—the head office of the Royal Dutch Shell oil company in The Hague. As always, the atmosphere was tense. Had he not explicitly been instructed to attend, Plesman would not have come. Across the table, Fokker knew this might well be his final opportunity to get his foot in the door of KLM's aircraft acquisition policy. Now that the government needed his company's military aircraft, he finally had a lever with which to pry open that door again, even though Plesman had taken such care to nail it shut. Fokker asked Plesman to consider his Design 175, for an eighteen-passenger, twin-engined, high-wing airliner, very similar in appearance to the Douglas DC-5, which KLM was then considering for its domestic routes. Design 175 failed to raise Plesman's interest, and he also brushed aside Design 177, a three-engined, high-performance aircraft for the European routes. The days of three-engined airplanes had passed, he said.

The middle engine restricted the pilot's vision, could smear the wind-shield with oil, and presented extra fire hazards because fuel lines had to be built into the fuselage.

On the other hand, Plesman was prepared to discuss Design 173—a four-engined, long-range airliner with engines in tandem configuration and equipped with a pressurized cabin for twenty-four passengers—if Fokker could give him a reliable indication of price and delivery dates. Fokker could not do so at that point, and the meeting broke up without an agreement.[57] But at least they were talking, and Fokker was intent on using the pressures of the times to push the balance in his favor. Progress was difficult, however, marred by the mutual distrust between the two men. Talking to newspaper reporters on March 14, Fokker failed to con-ceal his intense dislike for Plesman and the way he had contributed to the demise of the Nederlandsche Vliegtuigenfabriek. Still, in view of the on-going talks, he was genuinely embarrassed when he found his opinions in print the next day under the headline, "Fokker on Plesman."[58] For once, the outburst went unanswered, and discussions continued between repre-sentatives from both sides, although very little progress was made. Satis-fied that an initial breakthrough had been achieved and the government was now receptive to his case, Fokker departed for the United States once again, in the company of his secretary, Jojo de Leuw.

In the course of further discussions in Holland, attention came to focus on Design 193—an improved version of Design 175—a twin-engined air-liner for the European routes, which Fokker and Plesman had discussed in March; and on Design 180. The latter project was the result of discus-sions between the Fokker company and the Holland-America shipping line in 1938 on the possibilities of building a pressurized airliner with transatlantic range. In its final form, Design 180 was to be a four-engined airplane with two tail booms, a short fuselage, and sleeping accommoda-tion for twenty-six passengers in the wings.[59] But with continuing fragile relations between the Fokker company and KLM, progress was difficult to achieve. KLM dragged its heels and made it clear to all concerned that they preferred to stick with American aircraft, no matter what the Fokker engineers were designing.

By the time KLM finally wrenched permission from the government to order four DC-5's, its plans for medium- and long-range aircraft were shifting to Lockheed's drawing-board projects: the L-44 Excalibur and the L-49 Constellation. In May the government installed an interde-partmental advisory commission to monitor discussions on KLM's procurement of Fokker aircraft. Two months later, on July 6, 1939, the government also took the lead in setting up a special contact commission to mediate between KLM and Fokker. Evidently, some provision for the Fokker designs had to be made. Still, Plesman managed to obtain permission from the government to order six long-range aircraft for the East Indies service (the choice was left free between the DC-4E and the Constellation) and four Lockheed Excaliburs for the European network. That permission came at the price of having to order four Fokker F.24's (the final version of Design 193) and one Fokker F.180 Intercontinental, for which the government would pay subsidies to both parties. As a result of the complexities that this arrangement introduced, negotiations on the F.24 contract dragged on until January 22 and 23, 1940, when final agreement was reached between the government and the Fokker company on the construction of four F.24's, which would be made available to KLM at a reduced price.[60]

None were built. The German invasion and occupation of Holland brought a violent end to the development of these aircraft. Work on the F.24 was discontinued in November 1940, and the F.180 design was kept afloat until the German authorities ordered the project's cessation in the summer of 1941.[61]

## THE END OF THE ROAD

Tony Fokker did not live to see the outcome of his efforts to recapture lost ground in civil aircraft production. In April 1939 he left Holland for the last time, gratified that his efforts to pressure KLM into accepting his company's latest designs were moving in the right direction. Life was definitely taking a turn for the better. An affair was burgeoning between him and his personal secretary, Jojo de Leuw, the outcome of their winter's sojourn in Saint Moritz. In June she followed him to his home in Nyack, where they would be able to enjoy each other's company undisturbed. The petite blonde brought happiness and renewed energy to the now asthmatic forty-nine-year-old Fokker, as well as bringing flowers to his Undercliff Manor.[62] There, at the end of a dead-end street, she gave him the incentive to pick up the pieces after the destruction of his beloved *Q.E.D.* and start planning for a new ship and new horizons. But it was not to be.

On December 1, 1939, still plagued by sinusitis and asthmatic complaints, Fokker went to see an osteopath to have an obstruction removed from his nose. The operation went tragically amiss. Under general anesthesia, he failed to regain full consciousness and had to be rushed to New York's Murray Hill Hospital in a semicomatose condition. Afterward it was established that he was suffering from pneumococcus meningitis, which may have been present before his nose surgery and contributed to its disastrous consequences. For three weeks he remained in the hospital, slowly improving at first, until he was again able to reply to a few questions; but after two weeks his fever suddenly rose, and he relapsed into a coma. To comfort Jojo, their mutual close friend, Wim van Neijenhoff, hurried across the Atlantic. Their bedside visits soon turned into a death wake. On December 21 his doctor, Robert Cushing, described his condition as critical. Tony Fokker was losing his final battle, and though his doctors tried various recently developed antibacterial drugs and blood transfusions, he did not respond to treatment, and the infection spread. In the early morning hours of December 23, 1939, around half past six, he passed away.[63]

Fokker's body was returned to Undercliff Manor. There, at the end of the road in the shelter of the rocks, his coffin was placed in the hallway he had liked so much, surrounded by flowers. Jojo and his friends gathered at the house at the end of the street on December 26 for a private farewell service conducted by a Reverend Mitchell, who had become a friend in recent years when Fokker had shown himself more receptive to religion. After the service, the body in its his rose-bedecked coffin was carried away to be cremated. Jojo de Leuw returned to Holland to be comforted by her family. Van Nijenhoff followed, with Fokker's ashes, after formalities had been seen to, arriving in Amsterdam on January 29. Five days later, on February 3, 1940, the ashes were interred at the Westerveld Cemetery near Haarlem in a ceremony attended by some 1,500 people. His mother had helped to make the arrangements despite her old age; it was the last and only thing she could do for him.[64] Among the black-clad mourners assembled around the coffin containing the urn, Jojo stood out: a frail blonde girl in a light coat, trapped in her silent grief.[65]

On April 19, 1941, the Fokker company dedicated a monument on the grave of its founder: an octagonal pillar on top of which a bronze bird spread its wings, ready to fly away. Under those wings, Anthony Fokker rests.

# EPILOGUE

One of the many surprisingly sincere statements that Anthony Fokker made in *Flying Dutchman* read: "My life has been paced by the airplane. Hurtling through space on what now seems a predestined course, I had no idea what that course was. Most of the time I merely hung on." At first glance, such an alarming denial of direction in his own life strikes the reader as less than true. Yet, on further thought, Fokker's admission holds several penetrating implications.

That Fokker's life was shaped by the airplane is undeniable. The lives of many early pioneers of aviation were shaped by it, and not infrequently, their lives were cut short by the very element that shaped them. Not all of their names have the ring that the name Fokker enjoys to this day. The role of the airplane in Fokker's life could have come about only because of the particular and privileged circumstances of his background. Family members were able to invest large sums of money in what must have appeared a lost cause at the time. That initial capital was vital to Fokker's later career. His practical genius in coming to grips with how to build an airplane would not, in itself, have been enough to bring him to the fore of the aeronautical community. Where others had to give up as they discovered that their real wings did not equal their imaginary ones, Fokker was able to hold out until luck came his way.

The incidental combination of capital, brilliance, a copied plane (the

Morane), German military aviation policy, and, above all, the outbreak of the war in August 1914 determined his career. Even then, Fokker never really gained control of his business. The fortunes of his German, Dutch, and American companies were not so much the result of careful planning or entrepreneurial strategy as of factors beyond his direct control. The war brought possibilities for growth that would have been undreamt of under peaceful conditions. Fokker was lucky he and his engineers were on the brink of solving the gun synchronization problem when the German military came knocking. Repeated design flaws in new aircraft and poor quality control did not prove detrimental in wartime conditions, and thus his company was able to continue aircraft development, which culminated in the D.VII, Germany's top-rate fighter aircraft of 1918.

With his reputation as a leading aircraft constructor firmly established, and presented with a viable design for an efficient airliner by the Schwerin development team just after the war, Fokker was well placed to take a prominent position in postwar aviation. His Dutch company, Nederlandsche Vliegtuigenfabriek, developed favorably in the 1920s. But initial commercial successes resulting from adapting and improving the Fokker design concept, enabling transatlantic expansion of his business interests in the shape of Atlantic Aircraft and the American Fokker Aircraft Corporation, blinded Fokker to the need for continuous investment in research and development of aircraft embodying emerging new technologies.

As late as 1930, Fokker's perspective on aircraft development continued to be short-sighted in a competitive environment that increasingly demanded a long-term view and sizable investments, enabling radically new approaches to aircraft design and construction. The technological transformation in the aircraft industry from building a relatively small series of customized aircraft to mass-producing all-metal airplanes of standardized design set the stage for his downfall as a constructor and of his business enterprises. Two years later, in 1932, by a bizarre twist of fate, Fokker found himself forced to repeat the mistakes of the late 1920s that had cost him his leading position in the United States. He was obliged to build the giant F.XXXVI, which he did not believe in, at the cost of developing a viable airliner that might have taken his mixed-construction technology into the 1940s. The project nearly led to the demise of his Amsterdam factory, a process arrested only by the European rearmament drive of the late 1930s.

But Tony Fokker personified the ongoing changes in the global aviation industry in a different way. By the mid-1930s he had largely withdrawn from direct involvement in aircraft development to focus his attention on selling American-designed planes in Europe. In this capacity, Fokker was very successful, and his achievements were crucial in bringing about the rapid acceptance and market penetration of American-designed aircraft in Europe. In his somewhat obscure activities on behalf of Douglas and Lockheed lay Fokker's longest-lasting contribution to the development of aviation. More important than the trend-setting trimotor designs of the 1920s that bore his name was Fokker's marketing of the latest U.S.-manufactured aircraft, which facilitated their market penetration in Europe. By 1939 the American aircraft industry had acquired a position that led the way to a worldwide dominance in aircraft construction that exists to this day.

What kind of a person was Tony Fokker? The dream of any biographer is to be able to sit down with his subject and, at the end of the book, ask the vital questions: Is all of this true? Was this how it really happened? What about . . . ? In the case of Fokker, drawing firm conclusions from the fragmentary source materials that survived the teeth of time was no easy business. The picture that emerges from this book may well have been colored by the slightly larger-than-life memories of the various people who wrote down their recollections and interpretations of the man. Because few primary sources have survived, the use of various published and unpublished memoirs increased the element of uncertainty common to all history books.

Certainly, Fokker was a difficult person to get along with. His posture as a self-styled engineer made him demanding and obstinate. Money meant little to him, and his uncompromising eccentricity wore out even his closest friends. There was no regularity whatsoever in his daily activities, no recognizable pattern in which he tackled everyday chores. Half the time he was away on business or pleasure trips, his course endlessly meandering across the borders of Europe, and seldom was he on time for an appointment. In his later years, his restlessness grew worse. He seemed to prefer to wake up in hotel rooms, aboard railroad sleeping cars, or on ships. Not even those who loved him most could keep up with him. No amount of affection seemed to be strong enough to compensate for his erratic behavior and egotism. Fokker's first wife, Tetta, left him through divorce; his second wife, Violet, vented her desperation with even more crushing finality. It took him ten years to recover from that blow and find a renewed sense of belonging in the last six months of his life.

Did Anthony Fokker die before his time? By average standards his life was short. Nonetheless, it had been complete, full of unique developments that allowed this rich man's son an extraordinarily uncompromising pursuit of his winged dream and made him into one of the figures who stand out in the history of flight. He would not have been able to add much more to that history, for at the time of his unexpected death in December 1939, the pioneer days of practical, Fokker-style engineering had come to pass. Although his name lives on, the company that bore his signature, Fokker Aircraft, collapsed in 1996.

```
┌─────────────────────────────────┐
│                                 │
│            NOTES                │
│                                 │
└─────────────────────────────────┘
```

## CHAPTER 1. GROWING WINGS

1. For the Fokker genealogy, see A. A. Vorsterman van Oyen, *Stam- en Wapenboek der Aanzienlijke Nederlandsche Familiën,* vol. 1 (Groningen: Wolters, 1885), 284–286.

2. Lou de Jong, *Het Koninkrijk der Nederlanden in de Tweede Wereldoorlog,* vol. 11a, *Nederlands-Indië—I, Eerste Helft* (The Hague: Staatsuitgeverij, 1984), 55–56.

3. Memo, Haarlem Municipal Civil Registration Department, December 17, 1918: Haarlem Municipal Archive, new archive, 126-1.

4. In the autobiography written for Tony Fokker by the *New York Evening Post* aviation reporter, Bruce Gould, Fokker recalled that "most of the residents [of Haarlem], like my father, were retired planters or business men living out the final years of their lives sedately." Anthony H. G. Fokker and Bruce Gould, *Flying Dutchman: The Life of Anthony Fokker* (New York: Henry Holt, 1931), 17.

5. Fokker and Gould, *Flying Dutchman,* 21.

6. Ibid., 15.

7. Anthony Fokker, *HBS Feestblad* (Haarlem), September 4, 1924.

8. Fokker and Gould, *Flying Dutchman,* 22–24.

9. Photocopy of certified transcript of A. H. G. Fokker's school results at Haarlem's Hoogere Burger School (HBS-B), October 21, 1955: Aeronautical Collection, Thijs Postma, Hoofddorp, Hegener Papers, dossier of Fokker material.

10. Jacob T. Cremer later became the Dutch ambassador to the United States (1918–1920): Memorandum for the American minister by the representative of

the War Trade Board at The Hague regarding the Deli Maatschappij, Amsterdam, prepared by the commercial attaché to the U.S. Embassy in The Hague, Edwards, 29/3/19: U.S. National Archives, Washington, D.C., State Department, Record Group 84, Netherlands Post File, General Correspondence, 1919, 711.3.

11. Correspondence regarding Frits Cremer's school problems: Algemeen Rijks Archief (Netherlands National Archives), The Hague, Collection 2.21.043, Cremer Family Papers, no. 204.

12. Frits Cremer to his father, received February 1 and March 18, 1910: Algemeenrijks Archief, 2.21.043, Cremer Family Papers, no. 205.

13. Fokker and Gould, *Flying Dutchman,* 26–32.

14. Frits Cremer to his father, April 26, 1910: Algemeen Rijks Archief, 2.21.043, Cremer Family Papers, no. 205.

15. Anthony Fokker to his mother, undated (March 1910): Aviodome Archive, Schiphol, Anthony Fokker dossier, correspondence.

16. Tony Fokker to his mother, undated (May 1910): Aviodome Archive, Anthony Fokker dossier, correspondence.

17. Tony Fokker to his mother, May 9, 1910: Aviodome Archive, Anthony Fokker dossier, correspondence.

18. Felix Ph. Ingold, *Literatur und Aviatik: Europäische Flugdichtung, 1909–1927* (Basel: Birkhäuser Verlag, 1978), 19–27.

19. Ibid., 28–33.

20. On the rise of the German airship, see Henry Cord Meyer, *Airshipmen, Businessmen, and Politics, 1890–1940* (Washington, D.C.: Smithsonian Institution Press, 1991), 21–121. In a wider context, the phenomenon is captivatingly described in Peter Fritsche, *A Nation of Fliers: German Aviation and the Popular Imagination* (Cambridge: Harvard University Press, 1992), 9–58.

21. C. C. Bergius, *Die Strasse der Piloten: Die abenteuerliche Geschichte der Luftfahrt* (München: Wilhelm Heyne Verlag, 1971), 284–86.

22. Ibid., 287–93.

23. Robert Wohl, *A Passion for Wings: Aviation and the Western Imagination, 1908–1918* (New Haven: Yale University Press, 1994), 100. In this excellent book, Wohl paints a powerful image of the exploits of the early aviators and their impact on society and culture.

24. Herman Fokker to Henri Hegener, April 22, 1923: Aeronautical Collection, Thijs Postma, Hegener Papers.

25. Fokker and Gould, *Flying Dutchman,* 37.

26. Tony Fokker to his mother, July 30, 1910: Aviodome Archive, Anthony Fokker dossier, correspondence.

27. Fokker and Gould, *Flying Dutchman,* 10.

28. Fokker to his mother, July 30, 1910: Aviodome Archive, Anthony Fokker dossier, correspondence. Because Tony Fokker had signed the registration form for the Technikum himself, he suggested that his father might use his, Tony's, legal status as a minor to get a refund.

29. Tony Fokker to his parents, October 12, 1910: Aviodome Archive, Anthony Fokker dossier, correspondence.

30. Fokker and Gould, *Flying Dutchman,* 38–43; Henri Hegener, *Fokker: The Man and the Aircraft* (Letchworth: Harleyford Publications, 1961), 14; A. R. Weyl, *Fokker: The Creative Years,* 2d ed. (London: Putnam, 1985), 10–12.

31. Bergius, *Die Strasse der Piloten,* 327–30.

32. Weyl mentions that photographs of the Fokker–Von Daum plane were published in *Flugsport* on November 2, 1910, and that earlier photos show the plane standing between the two Fachschule biplanes. Weyl, *Fokker,* 12–15.

33. On the airship as a symbol of popular nationalism in this period, see Fritsche, *Nation of Fliers,* 9–43.

34. Fokker to his mother, dated April 1911: Aeronautical Collection, Thijs Postma, Hegener Papers.

35. Fokker and Gould, *Flying Dutchman,* 52–53; Weyl, *Fokker,* 15–17.

36. A photograph of his license is kept in the Aviodome Archive at Schiphol Airport, Amsterdam.

37. Hans Karl Rehm, "Der fliegende Holländer, die Wahrheit im Fall Fokker," *Mainzner Anzeiger,* July 21, 1933; Kurt Ries, "Määnzer Gefladder: Eine kleine Geschichte der Fliegerei in und um Mainz," *Mainzer Vierteljahresheft für Kultur, Politik, Wirtschaft und Geschichte* 2 (1982), no. 3: 114–39.

38. Fokker and Gould, *Flying Dutchman,* 46, 54–56.

39. Ibid., 57–61.

40. Herman Fokker to Anthony, September 30, 1912: Aviodome Archive, Anthony Fokker dossier, correspondence.

41. Fokker to his father, November 29, 1911: Aviodome Archive, Anthony Fokker dossier, correspondence.

42. For the background of this phenomenon, see Fritz Fischer, *Griff nach der Weltmacht: Die Kriegszielpolitik des kaiserlichen Deutschland, 1914–18* (Düsseldorf: Droste Verlag, 1961), 22–36.

43. John H. Morrow Jr., *The Great War in the Air: Military Aviation from 1909 to 1921* (Washington, D.C.: Smithsonian Institution Press, 1993), 19–20.

44. Fokker and Gould, *Flying Dutchman,* 70. The partnership with Haller is explained in Weyl, *Fokker,* 26–31.

45. Fokker to his father, undated (first week of April 1912): Aeronautical Collection, Thijs Postma, Hegener Papers. The translation follows Fokker's original choice of words and punctuation as closely as possible in English.

46. Hegener, *Fokker,* 18.

47. Fokker and Gould, *Flying Dutchman,* 86–91; Weyl, *Fokker,* 30–31.

48. Fokker to his father, May 3, 1912: Aviodome Archive, Anthony Fokker dossier, correspondence; Fokker and Gould, *Flying Dutchman,* 85.

49. Herman Fokker to Anthony, May 24, 1912: Aviodome Archive, Anthony Fokker dossier, correspondence.

50. Fokker and Gould, *Flying Dutchman,* 92–95.

51. Ibid., 70.

52. A. H. G. Fokker and Bruce Gould, *De Vliegende Hollander* (Amsterdam: Van Holkema en Warendorf, 1931), 108. It is most peculiar that the episode with Ljuba Galantschikova did not find its way into the original 1931 American

edition of *Flying Dutchman* but *was* recounted in the subsequent Dutch and German editions of the book, suggesting that Fokker, while going over the Dutch proofs, inserted these personal revelations as an afterthought. In the Dutch (and German) editions, the affair with Ljuba Galantschikova overshadows the bid for the czarist army contract in a chapter titled "De Puschka"/ "Die Puschka" (The ace). Fokker and Gould, *Vliegende Hollander,* 104–11; A. H. G. Fokker and Bruce Gould, *Der fliegende Holländer: Das Leben des Fliegers und Flugzeugkonstrukteurs A. H. G. Fokker* (Zürich: Rascher & Cie. Verlag, 1933), 112–20.

53. Fokker and Gould, *Vliegende Hollander,* 108, 110.

54. Ibid., 110.

55. Reminiscences of Friedrich Seekatz, who first met Fokker at Johannisthal in the fall of 1912 and was employed by him from August 1914. In a letter to Marius Beeling (then retired director of the postwar Fokker factory in Amsterdam), Seekatz hinted at Fokker's suicidal tendencies. Seekatz to Beeling, February 28, 1969: Fokker Archive, Beeling Papers, miscellaneous correspondence. Fokker and Galanschikova were to meet seventeen years later, in July 1930, in New York (see chapter 7).

56. Fokker and Gould, *Vliegende Hollander,* 111. True enough, among the few surviving papers in the Aviodome Archive from Fokker's early years, there are several gaily written postcards addressed to him by female members of his staff that suggest more than a purely employer-employee relationship.

57. Fokker and Gould, *Flying Dutchman,* 97–101.

58. Herman Fokker to Anthony, September 30, 1912: Aviodome Archive, Anthony Fokker dossier, correspondence (Herman Fokker's emphasis).

59. Photo/postcard, Tony Fokker to his mother, November 24, 1912: Aviodome Archive, Anthony Fokker dossier, correspondence.

60. C. C. Küpfer, typescript for a commemoration volume for the Fokker Aircraft Company's fortieth anniversary, p. 13: Fokker Archive (Verhoef) (hereafter cited as Küpfer typescript). Küpfer was the company's secretary from 1948 to 1959 and apparently based his typescript on source material that has since been lost. Weyl, *Fokker,* 35; Hegener, *Fokker,* 22.

61. Hegener, *Fokker,* 22. Küpfer typescript, 13.

62. John H. Morrow Jr., *German Air Power in World War I* (Lincoln: University of Nebraska Press, 1982), 7.

63. Kenneth Munson, *Historische Vliegtuigen, 1903–1914* (Amsterdam: Moussault, 1970), 130–31; Morrow, *Great War,* 37–38. I am indebted to Peter M. Grosz of Princeton, New Jersey, for additional background information.

64. Fischer, *Griff nach der Weltmacht,* 48ff.

65. Fokker and Gould, *Flying Dutchman,* 113.

66. On the acquisition of the Morane-Saulnier and the development of the new Fokker M.5 plane, see Weyl, *Fokker,* 61–82.

67. Weyl, *Fokker,* 78, 81–84.

68. Founding statute of Fokker Aeroplanbau GmbH, Schwerin-i.M., July 7, 1914: Deutsches Zentral Archiv (Potsdam), no. 87.40, dossier 62, Kriegsverband Deutsche Flugzeugproduzenten.

## CHAPTER 2. THE FORTUNES OF WAR

1. Fritsche, *Nation of Fliers*, 43–58.
2. Morrow, *Great War*, 7–11, 33–37. Morrow suggests that the development of a standard reconnaissance plane focused the efforts of the German aircraft industry "perhaps too narrowly" on a single type (54).
3. Ibid., 35, 39, 45, 47, 57. Morrow quotes the strength of the Royal Flying Corps early in 1915 at only 85–90 frontline aircraft, 106 in June, and 153 in September (113).
4. Ibid., 45.
5. Ibid., 64.
6. Ibid., 71.
7. Fokker and Gould, *Flying Dutchman*, 117–18; Weyl, *Fokker*, 83–85; Morrow, *Great War*, 73. See also R. J. Castendijk, "A. H. G. Fokker," *Het Vliegveld* 1, no. 2 (1917): 28–29.
8. Reminiscences of Anthony Fokker by Friedrich Seekatz, February 25, 1969: Fokker Archive, Beeling Papers, miscellaneous correspondence.
9. Morrow, *German Air Power*, 25–26, 47. In October 1914 a contemplated takeover of an aircraft construction facility near Berlin came to naught. Hugo Hamacher to Tony Fokker, October 30, 1914: Fokker Archive, A. H. G. Fokker, box 2.
10. Fokker production statistics were compiled by Peter M. Grosz from authentic acceptance and forwarding data sheets: Grosz Archive. The factory efficiency of five major German aircraft producers in 1914–15 was as follows:

| Company | Year | Work Force | Aircraft Built | Planes per Worker |
|---|---|---|---|---|
| LVG | 1914 | 450 | 300 | 0.66 |
| LVG | 1915 | 920 | 1,020 | 1.11 |
| Aviatik | 1914 | 220 | 165 | 0.75 |
| Aviatik | 1915 | 330 | 316 | 0.95 |
| Albatros | 1914 | 745 | 336 | 0.45 |
| Albatros (Johannisthal) | 1915 | 1,235 | 851 | 0.69 |
| Fokker | 1914 | 110 | 32 | 0.29 |
| Fokker | 1915 | 480 | 260 | 0.54 |
| Rumpler | 1914 | 630 | 108 | 0.17 |
| Rumpler | 1915 | 835 | 210 | 0.25 |

The table was adapted from Morrow, *German Air Power*, 47.

11. Winfried and Walter Suwelack, *Josef Suwelack und der Traum vom Fliegen: Leben und Tod des Billerbecker Flugpioniers und Weltrekordlers Josef Suwelack, 30. April 1888 bis 13. September 1915* (Warendorf: Karl Darpe Verlag, 1988), 113. Suwelack's letter is quoted in Jörg Armin Kranzhof, *Fokker Dr.I,* in *Flugzeuge die Geschichte machten* (Stuttgart: Motorbuch Verlag, 1994), 11.

12. Tony Fokker to Grossherzoglich Mecklenburgisches Contingent-Kommando, December 22, 1914: Fokker Archive, Beeling Papers. See also

Fokker and Gould, *Flying Dutchman,* 120. On the financial problems of the German aircraft industry in 1914, see Morrow, *Great War,* 73–74.

13. Weyl, *Fokker,* 131.

14. Quoted in Hegener, *Fokker,* 2; idem, Lieutenant Parschau to Fokker, May 26, 1915: Fokker Archive, Beeling Papers, miscellaneous correspondence.

15. Tony Fokker to Grossherzoglich Mecklenburgische Contingent-Kommando, December 22, 1914: Fokker Archive, Beeling Papers, miscellaneous correspondence.

16. Correspondence between Anthony Fokker and his legal council in the naturalization dispute, Professor A. S. Oppenheim, Oppenheim to Fokker, February 22, 1919: Haarlem Municipal Archive (Holland), New Archive, no. 126-1.

17. Weyl, *Fokker,* 393–402.

18. Morrow, *Great War,* 90–92.

19. Alex Imrie, *The Fokker Triplane* (London: Arms & Armour Press, 1992), 11. See also Kranzhof, *Fokker Dr.I,* 40–43, 46.

20. Fokker-Schneider legal file, 1916–26: Fokker Archive, V/Alg. Secr., no. 67.

21. Kranzhof describes the actual functioning of the synchronization mechanism in some detail in *Fokker Dr.I,* 49–53.

22. Fokker and Gould, *Flying Dutchman,* 122–27. See also Hegener, *Fokker,* 23–26; Kranzhof, *Fokker Dr.I,* 43–46; Weyl, *Fokker,* 95–97. Weyl strongly disclaims here, as elsewhere in the book, the direct involvement of Fokker in the engineering of the synchronization mechanism and gives all credit for the invention to Lübbe and his assistants, Heber and Leimberger. In doing so Weyl takes the case well beyond the limits of credibility. There is no reason to believe Fokker was not personally involved in this very practical engineering process, especially since he excelled in hands-on solutions. The invention was in all likelihood the outcome of a team effort, for which Fokker, as employer, could rightfully claim intellectual ownership, even though Lübbe might have done most of the work (as Kranzhof suggests).

23. In his autobiography, Fokker claimed that he was actually ordered to go and shoot down an Allied plane (Fokker and Gould, *Flying Dutchman,* 131–38), but Weyl convincingly rejects the story as fictitious in *Fokker* (100–2).

24. Fokker and Gould, *Flying Dutchman,* 134–35.

25. Ibid., 135–36.

26. Morrow, *German Air Power,* 20–22, 56–59, 80–81.

27. Fokker to Königliche Inspektion der Fliegertruppen (Idflieg), September 28, 1916: Fokker Archive, A. H. G. Fokker, box 1. See also Fokker and Gould, *Flying Dutchman,* 179–84. Fokker claimed to have produced as many as 6,000 gun gears per month at one time, which must be hugely exaggerated, since it would have made no sense to produce six times the number of synchronization mechanisms of the total airplane production in Germany, which failed to meet its aim of building 1,000 planes a month until 1918. Morrow, *Great War,* 112, 162–65, 227, 305.

28. Weyl, *Fokker,* 190–91.

29. Fokker and Gould, *Flying Dutchman,* 184–89; Hegener, *Fokker,* 32; Weyl, *Fokker,* 189.

30. Correspondence, 1916–26, in the file on the Fokker-Schneider patent case: Fokker Archive, V/Alg. Secr., no. 67. See also Kranzhof, *Fokker Dr.I,* 57–58.

31. Correspondence, 1916–26, in the file on the Fokker-Schneider patent case: Fokker Archive, V/Alg. Secr., no. 67.

32. Weyl, *Fokker,* 105–6, gives the estimated top speed of the E.I as 80 mph. See also Morrow, *Great War,* 92.

33. Acceptance figures for Fokker aircraft, 1914–18, courtesy of Peter M. Grosz. See also Morrow, *Great War,* 106. For a recent example of the exaggerated emphasis on the Fokker scourge, see Wim Wennekes, "Tonny Fokker (1890–1939), 'Zolang ik iets heb om voor te vechten voel ik mij lekker,'" in *De aartsvaders: Grondleggers van het Nederlandse bedrijfsleven* (Amsterdam: Uitgeverij Atlas, 1993), 393–422.

34. Max Immelmann to Anthony Fokker, August 5, 1915: Fokker Archive, Beeling Papers, miscellaneous correspondence. See also Kranzhof, *Fokker Dr.I,* 14. Production of the E.I–E.IV totaled some 415 aircraft.

35. Morrow, *German Air Power,* 41; Imrie, *Fokker Triplane,* 13; Fritsche, *Nation of Fliers,* 69, 74–100.

36. Morrow, *German Air Power,* 90–92.

37. Fokker and Gould, *Flying Dutchman,* 142–43, 148–50. See also Weyl, *Fokker,* 113–25.

38. Fokker and Gould, *Flying Dutchman,* 148; Weyl, *Fokker,* 147–91, 217.

39. Peter M. Grosz, George Haddow, and Peter Schiemer, *Austro-Hungarian Army Aircraft of World War One* (Mountain View, Calif.: Flying Machine Press, 1994), 339–40.

40. Seekatz to Fokker, October 1, 1917: Aviodome Archive, Fokker-Germany dossier.

41. Tony Fokker to his father, October 10, 1917: Aviodome Archive, Anthony Fokker dossier, correspondence.

42. Fokker-Werke mbH, Schwerin, graph of male and female work force since October 1917; *Lehrplan der Fokker-Werke, Schwerin,* undated, apparently 1917: Aeronautical Collection, Peter M. Grosz.

43. Weyl, *Fokker,* 147–69.

44. Peter M. Grosz, "Reinhold Platz and the Fokker Company," *Over the Front* 5, no. 3 (1990): 213–20. See also Imrie, *Fokker Triplane,* 15–21.

45. Fokker to his father, October 10, 1917: Aviodome Archive, Anthony Fokker dossier, correspondence.

46. Diary of Hendrik Walaardt Sacré, commander of the Netherlands Air Corps, entries of August 8 and 27, 1917: Royal Netherlands Air Force, Air Historical Branch, The Hague; Netherlands military attaché in Berlin, Muller Massis, to Cornelis Snijders (the Netherlands chief of staff), December 24, 1917: Archive of the Netherlands Ministry of Defense, The Hague, General Headquarters, dossier no. GG.151b, 240. See also R. A. F. Hezemans, "To Fly or Not to Fly, That's the Question," in A. P. de Jong, ed., *Vlucht door de tijd: 75 jaar Nederlandse luchtmacht* (Houten: Unieboek, 1988), 40.

47. Consul-General Wolff of the Netherlands Legation in Berlin to Hendrik Walaardt Sacré, commander of the Netherlands Air Corps, January 15, 1918:

Archive of the Netherlands Ministry of Defense, General Headquarters, dossier no. GG.151b, 240.

48. Bernard Plage, director of the Fokker Works in Schwerin, to the Netherlands minister of war, November 21, 1918; Von Geusau, Netherlands minister of war, to Plage, December 6, 1918: Archive of the Netherlands Ministry of Defense, Ministry of War, no. 2774, Afd. 4, no. 57; de Jong, *Vlucht door de tijd,* 419–20.

49. Morrow, *German Air Power,* 95–120.

50. Protocol of the negotiations between Tony Fokker and Hugo Junkers in Berlin and Dessau on December 16, 18, 22, 1916; Junkers's reply to the proposal by Fokker for remuneration of a license on the aircraft patent, February 1, 1917: Archive of the Deutsches Museum, Munich, Germany, Junkers Archive, 0201/T11.

51. Minutes of a meeting between Horter and Lottmann of the Junkers Werke, February 2, 1917: Tony Fokker to Hugo Junkers, April 16, 1917: Archive of the Deutsches Museum, Junkers Archive, 0201/T11.

52. Agreement between Tony Fokker and Hugo Junkers, June 19, 1917: Archive of the Deutsches Museum, Junkers Archive, 0201/T11.

53. Contract of foundation of the Junkers-Fokker Werke AG Metallflugzeugbau between Hugo Junkers and Anthony Fokker, August 20, 1917: Archive of the Deutsches Museum, Junkers Archive, 0201/T11

54. Notice on the negotiations with the Junkers-Fokker Werke AG by Idflieg, May 13, 1918: Archive of the Deutsches Museum, Junkers Archive, 0201/T9.

55. Weekly surveys *(Wochenberichte),* Junkers-Fokker Werke, January–December 1918: Fokker Archive, series KW, box 67; Peter M. Grosz and Gerard Terry, "The Way to the World's First All-Metal Fighter," *Air Enthusiast* 25 (August 1984): 60–76.

56. A. D. Fischer von Poturzyn, *Junkers und die Weltluftfahrt: Ein Beitrag zur Entstehungsgeschichte Deutscher Luftgeltung, 1909–1933* (München: Richard Pflaum Verlag, 1933), 34; Carl H. Pollog, *Hugo Junkers: Ein Leben als Erfinder und Pionier* (Dresden: Carl Reissner Verlag, 1930), 88–89.

57. Correspondence and records belonging to legal proceedings between Junkers and Fokker on the cantilever wing construction has survived the teeth of time in the Fokker Archive, series KW, boxes 22–26.

58. Günter Schmitt, *Hugo Junkers und seine Flugzeuge* (Stuttgart: Motorbuch Verlag, 1986), 35–39. See also Fokker and Gould, *Flying Dutchman,* 147–48.

59. Peter M. Grosz, "Archive: Celebrating the 70th Anniversary of the Fokker V.1," *World War I Aero* 113 (1987): 28–39. See also Peter M. Grosz, "Reinhold Platz and the Fokker Company," *Over the Front* 5, no. 3 (1990): 213–22, specifically 220. Grosz's assessments, based on newly uncovered documents, contrast sharply with Weyl's in *Fokker,* which offers a detailed technical description of how Weyl's hero, Platz, "designed" the cantilever wing (207–16).

60. Weyl, *Fokker,* 220–21. For a description of the Dr.I's structural principles, see Weyl, *Fokker,* 406–10.

61. Tony Fokker to his father, October 10, 1917: Aviodome Archive, Anthony Fokker dossier, correspondence.

62. Peter M. Grosz and A. E. Ferko, "The Fokker Dr.I: A Reappraisal," *Air Enthusiast* 8 (October 1978): 9–26; Imrie, *Fokker Triplane*, 23–25. Weyl claimed that engines were imported from the Swedish firm of Thulin, but there is no evidence to back this up. Weyl, *Fokker*, 218–23.

63. Imrie, *Fokker Triplane*, 27.

64. Ibid., 39–51, 114–15; Kranzhof, *Fokker Dr.I*, 70–89; Weyl, *Fokker*, 233–40. Interestingly enough, the issue of quality control in production was never really resolved during Fokker's career. In the famous Knute Rockne crash of March 31, 1931 (see chapter 7), which contributed to the demise of aircraft of mixed wood and metal construction, substandard workmanship and bad gluing were key issues.

65. Fokker and Gould, *Flying Dutchman*, 203.

66. On the friendship between Fokker and Von Richthofen, see Tony Fokker to his father, October 10, 1917: Aviodome Archive, Anthony Fokker dossier, correspondence.

67. Morrow, *German Air Power*, 79, 124; Weyl, *Fokker*, 268–84.

68. Weyl, *Fokker*, 286.

69. Peter M. Grosz, "Fokker's D.VIII . . . the Reluctant Razor," *Air Enthusiast* 17 (December 1981): 61–73 (quoted on 71).

70. Idflieg found that the wing spars of the series machines were underdimensioned compared to the prototype, which had undergone acceptance tests at Adlershof. Grosz, "Fokker's D.VIII," *Air Enthusiast* 17 (December 1981): 61–73. See also Weyl, *Fokker*, 328–41. In Fokker's own version of the matter, presented in *Flying Dutchman* (173–78), he blamed Idflieg for insisting on unnecessary strengthening of the rear wing spars, which led to uneven stress distribution inside the wings and thus caused the fractures. True enough, the reputation of the E.V was sufficiently tainted to preclude acquisition by the Dutch Air Corps of a number of the D.VIII's that Fokker railroaded into Holland in March 1919. The Dutch settled for the proven D.VII biplane design. De Jong, *Vlucht door de tijd*, 63–65.

71. Official French text of the Armistice Agreement of November 11, 1918, *Der Waffenstillstand 1918–1919: Das Dokumenten-Material der Waffenstillstands-verhandlungen von Compiègne, Spa, Trier und Brüssel, herausgegeben im Auftrag der Deutsche Waffenstillstands-Kommission* (Berlin 1928), band 1, 75 (translated by the author; emphasis added).

## CHAPTER 3. CORNERED AMID CHAOS

1. Fokker to his father, October 10, 1917: Aviodome Archive, Anthony Fokker dossier, correspondence.

2. Seekatz to Beeling, February 25, 1969: Fokker Archive, Beeling Papers, miscellaneous correspondence.

3. Ibid.

4. Fokker to his father, October 10, 1917: Aviodome Archive, Anthony Fokker dossier, correspondence.

5. The *Tetta* was a sizable yacht, offering sleeping accommodation to at least two adults and carrying 75 square meters of sail. Postcard, Tony Fokker to his cousin, Anthony Fokker, May 31, 1916: Aviodome Archive, Anthony Fokker dossier, correspondence.

6. Fokker and Gould, *Flying Dutchman,* 228.

7. Theodore von Kármán, *The Wind and Beyond: Pioneer in Aviation and Pathfinder in Space* (Boston: Little, Brown, 1967), 85.

8. Fokker and Gould, *Flying Dutchman,* 144.

9. Secretary Seglerhaus am Wannsee, B. Urbich, to Fokker, July 15, 1918: Fokker Archive, A. H. G. Fokker, box 2.

10. B. Urbich to Fokker, June 21, 1918: Fokker Archive, A. H. G. Fokker, box 2.

11. Fokker and Gould, *Flying Dutchman,* 208–10, 229–30.

12. Secretary Seglerhaus am Wannsee, B. Urbich, to A. H. G. Fokker, August 7, 1918; and Fokker to Urbich, June 25, 1918: Fokker Archive, A. H. G. Fokker, box 2.

13. For a balanced picture of the consequences of the military collapse on conditions in Germany, see such works as Fischer, *Griff nach der Weltmacht,* 820–56; and, more recently, Hans Mommsen, *Die verspielte Freiheit: Der Weg der Republik von Weimar in den Untergang, 1918 bis 1933* (Frankfurt am Main: Ullstein/Propyläen Verlag, 1990), 13–63.

14. Julius Stahn, secretary of the sailing club, to Fokker (addressed to Hotel Bristol, Berlin, with an indication in pencil that it was sent on to Holland), April 7, 1919: Fokker Archive, A. H. G. Fokker, box 2. Early in 1919, Stahn heard that Fokker was planning to have the ship transported to Holland, and implored him to sell it back to the club for the same price he had paid for it. Nevertheless, Fokker held on to the yacht and to his membership in the Wannsee club and was asked to pay his dues for keeping the *Erika* at Wannsee in June 1919, for which B. Urbich, then the club's secretary, sent him a bill in Amsterdam.

15. Julius Stahn to Fokker, October 28, 1918: Fokker Archive, A. H. G. Fokker, box 2.

16. Mommsen, *Die verspielte Freiheit,* 23–25.

17. Morrow, *German Air Power,* 139–40.

18. Fokker and Gould, *Flying Dutchman,* 218.

19. Ibid., 218–19.

20. Alex de Jonge, *The Weimar Chronicle: Prelude to Hitler* (London: Paddington Press, 1978), 31–34.

21. Seekatz to Beeling, February 25, 1969: Fokker Archive, Beeling Papers, miscellaneous correspondence.

22. Fokker and Gould, *Flying Dutchman,* 219–20.

23. Netherlands consul in Berlin to the mayor of Haarlem, Holland, December 14, 1918: Haarlem Municipal Archive, New Archive, dossier no. 126-1.

24. Travel permit in Fokker's German passport, issued by the Schwerin City Police Department on December 16, 1918: Aviodome Archive, Passports Anthony Fokker dossier.

25. Fokker and Gould, *Flying Dutchman*, 219.

26. Seekatz to Beeling, February 25, 1969: Fokker Archive, Beeling Papers, miscellaneous correspondence.

27. Plage to the Netherlands minister of war, November 21, 1918; and Netherlands minister of war to Plage, December 6, 1918: Archive Netherlands Ministry of Defense, Ministry of War files, no. 2774, Afd. 4, no. 57.

28. Weyl, *Fokker*, 356–57.

29. Fokker and Gould, *Flying Dutchman*, 222–23. Curiously enough, Fokker had forgotten the name of his shipping manager when digging in his memory for the book, and consequently the American version of the autobiography mentions a Wilhelm Hahn as the brains behind the scheme. In (photocopied) correspondence surviving in Holland from the original (pre–World War II) Fokker Archives, there is an incomplete exchange of letters between Heinrich Mahn and the management of the Dutch Fokker factory about Mahn's terms for helping Fokker with his memoirs. The idea was that Mahn would put his recollections on paper and send them to Fokker in New York, but agreement on the remuneration for his efforts was not reached. Heinrich Mahn to Fokker factory, August 16, 1930; and Fokker factory to Mahn, August 19, 1930: Aeronautical Collection, Thijs Postma, Hegener Papers.

30. Netherlands Post File, General Correspondence, 1919: U.S. National Archives, State Department, Record Group 84, Department of State, 711.3.

31. Memorandum by the American minister at The Hague on prohibition of export of securities from Germany, June 9, 1919: U.S. National Archives, State Department, Record Group 84, Netherlands Post File, General Correspondence 1919, 711.7.

32. British ambassador in The Hague, Robert Graham, to the British Foreign Office, May 4, 1920: British Public Record Office, London (Kew), Foreign Office Papers, FO 371, 4289.

33. As far as the German aircraft industry was concerned, parts of Rohrbach were moved to Denmark, Dornier to Switzerland and Italy, Junkers to the Soviet Union and Sweden, and Heinkel to Sweden and Finland. The phenomenon was not limited to the field of aircraft construction, but encompassed the whole realm of weapons production and the industries affiliated with it. For examples of industries moving to Holland, see Charles Verkuylen and Brandpunt Venlo, "De instrumentenfabriek 'Nedinsco' als centrum van internationale wapenhandel, 1921–1945," in *Venlo's Mozaïek: Hoofdstukken uit zeven eeuwen stadsgeschiedenis* (Maastricht: Limburgs Geschied- en Oudheidkundig Genootschap, 1990), 325–70. See also J. Enklaar, "Nederlandse hulp bij de illegale Duitse herbewapening na 1920," *Spiegel Historiael* 25 (1990): 178–84.

34. Lachsenberg, director of Junkers AG in Dessau, to Netherlands minister of finance, September 12, 1921: Archive of the Netherlands Ministry of Foreign Affairs, DEZ, 151–53.

35. Fokker and Gould, *Flying Dutchman*, 221–42.

36. Hermann Göring to Tony Fokker, July 12, 1919: Aeronautical Collection, Peter M. Grosz.

37. Hegener, *Fokker*, 36–38. Though Hegener does not use any notes in his

book, the remnants of his papers indicate that he carefully compiled his information from correspondence with a number of the people directly involved.

38. In an open letter to the editor of the journal *Windsock International* (vol. 3, no. 3, August 1987), aviation historian Peter M. Grosz quotes a document of 1920 by the Inter-Allied Aeronautical Control Commission that reported a shipment left for the consignee, Fokker Trompenburg Co., for the Dutch government on March 18, 1919.

39. Fokker and Gould, *Flying Dutchman,* 224–26. See also Minutes Joint Boards of Control of the Maatschappij tot Exploitatie der Staats-Spoorwegen and the Hollandsche Yzeren Spoorweg-Maatschappij, December 1918: Archive Nederlandse Spoorwegen, Utrecht, Notulen Raden van Commissarissen.

40. Unfortunately no records survive of the Oldenzaal customs checkpoint that go back as far as 1919, and files in the archive of the Netherlands Railways are sketchy in the extreme, which makes it impossible to trace how these rail transports were actually organized. Research in the archives of the Netherlands Ministry of Finance at the Netherlands National Archives (Algemeen Rijks Archief) in The Hague, and at the archive of the Netherlands Railways (Nederlandse Spoorwegen) in Utrecht, came up virtually blank.

41. Papers of the U.S. Legation in The Hague, 1919: U.S. National Archives, State Department, Record Group 84, 711.2.

42. United States Federal Bureau of Investigation Investigative Case Files, 1908–22, OG 349701, February 19, 1920: U.S. National Archives, FBI, Record Group 65, M1085, roll 00779.

43. Fokker and Gould, *Flying Dutchman,* 231–35. See also Tony Fokker in an interview with *The World,* New York, November 9, 1919.

44. Request for statement of citizenship (Netherlands consul in Berlin to mayor of Haarlem), December 14, 1918: Haarlem Municipal Archive, New Archive, 126-1; Netherlands Citizenship Law of 1892, art. 7, sec. 5, in J. C. Meijer, *Nederlandsche Staatswetten* (Sneek: Van Druten & Bleeker, 1893).

45. Netherlands consul in Berlin to the mayor of Haarlem, January 31, 1919: Haarlem Municipal Archive, New Archive, 126-1.

46. Telegram, mayor of Haarlem to Netherlands consul in Berlin, February 6, 1919; mayor of Haarlem to queen's commissioner for North Holland, March 10, 1919: Haarlem Municipal Archive, New Archive, 126-1.

47. Official marriage certificate, Anthony H. G. Fokker and Sophie M. E. von Morgen, March 25, 1919: Haarlem Town Hall, Civil Registration Department, 1919–94.

48. Photo from the collection of the author, March 25, 1919. See also Fokker and Gould, *Vliegende Hollander,* 271.

49. Heemskerk to Ruys de Beerenbrouck, March 25, 1919: Haarlem Municipal Archive, New Archive, 126-1.

50. Proof of Netherlands citizenship of Anthony Herman Gerard Fokker, April 24, 1919: Archive of the Netherlands Ministry of Foreign Affairs, The Hague, DEZ 151-44. Tetta von Morgen was specifically excluded from this citizenship settlement, even though article 5 of the Dutch Citizenship Law stipulated that the wife adopted the nationality of the man in marriage with a Dutchman.

51. "Ons plebisciet: De twintig groote mannen," *Het Leven,* December 11, 1922.

52. *De Vliegende Hollander* was a low-budget Corona Film production, directed by Gerard Rutten. The acting was poor, except for the lead, Ton Kuyl; the movie got very bad reviews from most critics and was a failure at the box office despite its subject matter. L. J. Jordaan, "'De Vliegende Hollander' vliegt niet hoog," *Vrij Nederland,* July 20, 1957, 9; Jan Blokker, "Ongeschikt voor film of strip," *de Volkskrant,* March 13, 1982.

53. *Persoonlijkheden in het Koninkrijk der Nederlanden in Woord en Beeld* (Amsterdam: Van Holkema & Warendorf, 1938), 482; *Biografisch Woordenboek van Nederland,* vol. 1 (The Hague: Nijhoff, 1979), 188–90.

54. Fokker and Gould, *Flying Dutchman,* 239–42.

55. After Tetta's death (as Mrs. Zehentner) in 1973, the executor of her will contacted the central office of the then Fokker-VFW combine in Düsseldorf, Germany, and offered the Gobelin for sale to surviving members of the Fokker family. It is unclear whether any transaction followed. Fridl Enzensberger to Fokker-VFW Zentralbüro, July 5, 1973: Fokker Archive, A. H. G. Fokker, box 2.

56. Correspondence, 1919: Fokker Archive, A. H. G. Fokker, box 1; Horter to SIW, January 5, 1920: Aviodome Archive, Fokker-Germany dossier, tax matters; miscellaneous correspondence, 1919–23: Fokker Archive, V/Alg. Secr., no. 67.

57. Schreiben der Schweriner Industrie Werke wegen Entschädigung, October 4, 1921: Staatsarchiv Schwerin (courtesy Peter M. Grosz).

58. Hermann Göring to Tony Fokker, July 12, 1919: Aeronautical Collection, Peter M. Grosz. Göring had been in Copenhagen since May 8, 1919. It is unclear whether Fokker even responded: the letter was simply stamped *Afleggen* (to be filed).

59. Seekatz to Beeling, February 25, 1969: Fokker Archive, Beeling Papers, miscellaneous correspondence. In November 1921 the number of employees had already dwindled to six, plus a work force of eight, including the cleaning ladies. SIW to Reichsverkehrsminister, November 12, 1921: Aeronautical Collection, Peter M. Grosz, Fokker file.

## CHAPTER 4. A BUSINESS OF SORTS

1. Lou de Jong, *Het Koninkrijk der Nederlanden in de Tweede Wereldoorlog,* vol. 1, *Voorspel* (The Hague: Staatsuitgeverij, 1969), 64ff.

2. J. J. Woltjer, *Recent verleden: Nederland in de twintigste eeuw,* 2d ed. (Amsterdam: Uitgeverij Maarten Muntinga, 1994), 17.

3. De Jong (ed.), *Vlucht door de tijd,* 419–20; J. F. van Dulm and F. C. van Oosten, *50 jaar Marine Luchtvaart Dienst, 1917–1967* (The Hague: Koninklijke Marine, 1967), 140.

4. Fokker and Gould, *Vliegende Hollander,* 272–85. In a letter to aviation journalist and Fokker biographer Henri Hegener, the deputy director of the factory from 1925 to 1935, Bruno Stephan, later confirmed that the story given in

the book is essentially correct, having been corrected from the version given in the original American publication for the better-informed Dutch readership. Stephan to Hegener, January 5, 1956: Aeronautical Collection, Thijs Postma, Hegener Papers, correspondence.

5. The original contract called for the delivery of twenty D.VII's for the Naval Air Arm, and seventy-two D.VII's and ninety-two C.I's for the Air Corps. Trompenburg was bought out for 3,674,000 guilders, and a new contract was drawn up between the Munitions Bureau and Fokker on April 28, 1920: Fokker Archive, Küpfer typescript, 47, 52.

6. F. Gerdessen, "De kleine luchtvaart in Nederland rond 1920," *Avia* 42, no. 1 (1983): 24–26; Peter van de Noort, "Luchtvaart rond 1920—een vervolg," *Avia* 42, no. 9 (1983): 318–19. The spasmodic activities of Fokker's Lucht-toerisme were discontinued in 1921.

7. Articles of foundation of the Naamlooze Vennootschap Nederlandsche Vliegtuigenfabriek of Amsterdam, July 21, 1919, Bijvoegsel tot de *Neder-landsche Staatscourant* 216, no. 1548 (October 9, 1919). Interestingly, the articles of foundation mentioned that Fokker was "living in Schwerin."

8. Articles of foundation of the Naamlooze Vennootschap Nederlandsche Vliegtuigenfabriek, art. 2.

9. Stephan to Hegener, August 30, 1958: Aeronautical Collection, Thijs Postma, Hegener Papers, correspondence.

10. Report of the Commission for the Advancement of the Aircraft Industry (Commissie Bevordering Vliegtuigbouw) on the Fokker Company, submitted to the minister of waterworks, June 21, 1935, p. 26: Archive of the Ministry of Transport and Waterworks, Civil Aviation Service, RLD, 14-III (cited hereafter as *Van Doorninck Report*).

11. One of the pictures taken at the ELTA was published in *Het Vliegveld* 3, no. 9 (1919): 206.

12. *Van Doorninck Report,* 27. The price he paid for the exhibition buildings was 800,000 guilders, plus annual leasehold charges of 18,000.

13. E. B. Wolff, D. Vreede, A. Plesman, and H. Walaardt Sacré, *Advies inzake steunverleening Nederlandsche Vliegtuigindustrie,* January 19, 1921: Algemeen Rijks Archief, Waterstaat, Staatscommissie, no. 21.

14. Stephan to Hegener, July 17, 1957: Aeronautical Collection, Thijs Postma, Hegener Papers, correspondence. Platz's successor, Marius Beeling, had similar views, describing him as a humorless man. Marius Beeling, "Herinneringen van oud-directeur en hoofdconstructeur Ir. Marius Beeling," *Avia* 18, no. 7 (1969): 326–29. Fokker let Platz go in 1931 because his technical knowledge had become completely outdated, but Platz kept believing in himself as a great constructor. As late as June 1955, he still thought himself on a par with modern technology, offering his help to the British de Havilland firm to straighten out problems with their revolutionary Comet jet airliner. Correspondence between Platz, A. R. Weyl, and Sir Geoffrey de Havilland, June 17, 24, 28, 1955: Aeronautical Collection, Peter M. Grosz, Platz-Weyl correspondence.

15. Minutes, eighth meeting of delegated board members of KLM, December 13, 1919: KLM Board Papers, series R-5.

16. Minutes, eleventh meeting of delegated board members of KLM, January 27, 1920: KLM Board Papers, series R-5.

17. Anthony Fokker to the director-general of the Postal, Telegraph, and Telephone Service (PTT), E. P. Westerveld, April 30, 1920: Archive Centrale Directie der PTT, Verbaal no. 7098H.

18. Fokker to the commander of the Air Corps, Col. Hendrik Walaardt Sacré, June 11, 1920; and Walaardt Sacré to the chief of the General Staff, June 19, 1920: Archive of the Netherlands Ministry of Defense, CLV, 39, 1876.

19. Miscellaneous correspondence in the file of the Fokker-Schneider case, 1916–33: Fokker Archive, V/Alg. Secr., no. 67.

20. M. Moslehner, Fokker's Berlin-based secretary, to Von Morgen, July 28, 1923: Fokker Archive, V/Alg. Secr., no. 67.

21. Hugo Alexander-Katz, Fokker's German patent lawyer, to Tony Fokker, November 16, 1923: Fokker Archive, V/Alg. Secr., no. 67.

22. Alexander-Katz to Fokker, July 16, 1926: Fokker Archive, V/Alg. Secr., no. 67.

23. Alexander-Katz to Fokker, December 16, 1926: Fokker Archive, V/Alg. Secr., no. 67.

24. Various correspondence, November–December 1926: Fokker Archive, V/Alg. Secr., no. 67.

25. Schneider memoirs (typescript), 153: Deutsches Museum, Munich, Germany, Franz Schneider Sammlung (courtesy Peter M. Grosz).

26. In the letter he claimed that his "representatives were later either influenced, or completely incompetent. In any case the adversaries went about the legal issue very well, while we did very poorly, and in a time so shortly after the revolution, when the circumstances were bad for me." Anthony Fokker to J. J. B. van der Mandere (of the J. J. B. van der Mandere, J. Lubbers, and E. Heldring law firm in Amsterdam), May 1, 1933: Fokker Archive, V/Alg. Secr., no. 67.

27. Archival boxes referring to this suit are stacked two feet high in the Fokker Archive: series KW, nos. 22, 23, 24, 25, 26.

28. Beschluss des Besitzsteueramt Schwerin, September 19, 1919: Stadtarchiv Schwerin (courtesy Peter M. Grosz).

29. Fokker and Gould, *Flying Dutchman,* 235–36.

30. Klage des Reichsmilitärfiskus, vertreten durch die Abwicklungsintendantur des Militär-Verkehrs-Wesens (Abteilung für Luftstreitkräfte) . . . wider die Schweriner Industrie Werke, December 2, 1920: Aviodome Archive, Fokker-Germany dossier, tax issues.

31. Ibid.

32. Ibid.

33. Werthauer, Engelbert, and Pröll, barristers in Berlin, to Landesgericht Schwerin, June 20, 1922: Aviodome Archive, Fokker-Germany dossier, tax issues.

34. Von Morgen to Fokker's Schwerin-based lawyer, Konrad Albrecht, January 29, 1923: Aviodome Archive, Fokker-Germany dossier, tax issues.

35. Reichsminister der Finanzen to Albrecht, September 20, 1926: Aviodome Archive, Fokker-Germany dossier, tax issues.

36. Report by J. A. Nederbragt on imports and exports between the Netherlands and the USSR in 1923, undated (January 1924): Archive of the Netherlands Ministry of Foreign Affairs, DEZ 87, Rusland. Küpfer, in his typescript (78), quotes a figure of 292 Fokker aircraft exported to the USSR between 1922 and the end of 1924 (50 D.VII's, 15 C.I's, 100 C.IV's, 126 D.XI's, and 1 D.XIII). On the basis of Russian sources, Alexander Boyd, *The Soviet Air Force since 1918* (London: Macdonald and Jane's, 1977), 8, quotes a grand total of about 500 planes.

37. Ben Knapen, *De lange weg naar Moskou: De Nederlandse relatie tot de Sovjet-Unie, 1917–1942* (Amsterdam: Elsevier, 1985), 98–103.

38. "'Hydravion,' de Nederlandsche vliegtuig-nijverheid van 1918–26," *Het Vliegveld* 11, no. 3 (1927): 88 (quoting Fokker statistics); Küpfer typescript, 78.

39. Director of the Central Bureau of Statistics (CBS) to finance minister, August 12, 1924: Archive of the Ministry of Finance, Secret Correspondence (Geheim Verbaal), 10216, no. 268.

40. Secretary-general of the Department of Justice to minister of finance, June 21, 1924: Archive of the Ministry of Finance, Secret Correspondence, 10216, no. 255; *Nieuwe Rotterdamsche Courant,* July 15, 1924; inspector for imports and duties, Amsterdam, to minister of finance, August 5, 1924: Archive of the Ministry of Finance, Secret Correspondence, 10217, no. 333.

41. G. Freund, *Unholy Alliance: Russian-German Relations from the Treaty of Brest-Litovsk to the Treaty of Berlin* (London, 1957), 1–140, 210–12; Edward L. Homze, *Arming the Luftwaffe: The Reich Air Ministry and the German Aircraft Industry, 1919–1939* (Lincoln: University of Nebraska Press, 1976), 1–18.

42. Homze, *Arming the Luftwaffe,* 9. See also Heinz J. Nowarra, *Die verbotenen Flugzeuge, 1921–1935: Die getarnte Luftwaffe* (Stuttgart: Motorbuch Verlag, 1980), 18–26.

43. Weyl, *Fokker,* 388–92, gives a fairly detailed, though rather suggestive picture of the closing of the secret deal with the Stinnes Concern, though he mistakenly links the affair with Fokker exports to the Soviet Union.

44. Boyd, *Soviet Air Force,* 8. See also Jean Roeder, *Bombenflugzeuge und Aufklärer: Von der Rumpler-Taube zur DO.23* (Bonn: Bernard & Graefe Verlag, 1990), 134–35; Küpfer typescript, 78.

45. Inspector of imports and duties, Amsterdam, to minister of finance, August 5, 1924: Archive of the Ministry of Finance, Secret Correspondence, 10217, no. 333.

46. Dossier, "Supply of Fokker Aircraft to Poland," November 1924: Archive of the Netherlands Ministry of Foreign Affairs, DEZ 151 *luchtvaart, bepalingen enz.*

47. Statistics of aircraft production at the Fokker factory compiled by a senior civil servant at the Ministry of Colonial Affairs, April 5, 1934: Archive of the Ministry of Finance, dossier no. 1330-7/33 (appendix to the *Van Doorninck Report*). In 1924 sales totaled 10,345,000 guilders. Küpfer typescript, 88. The statistics compiled by Colonial Affairs appear to be based on incomplete data.

Actual production and sales were probably higher still, but cannot be accurately established due to the incomplete state of the surviving records.

48. Netherlands envoy in Prague to Netherlands minister of foreign affairs, June 14, 1926: Archive of the Netherlands Ministry of Foreign Affairs, DEZ 151 *luchtvaart, bepalingen enz.*

49. The licensing agreement between Fokker and A. V. Roe was announced in the *New York Times* on October 21, 1928.

50. De Jong, *Vlucht door de tijd,* 420–21.

51. Part of a letter from Anthony Fokker to the commander of the Air Corps, quoted in a letter by the commander of the Air Corps to the chief of the General Staff, June 19, 1920: Archive of the Netherlands Ministry of Defense, Air Defense Command (CLV), dossier 39, no. 1876.

52. Duplicate of a letter from the minister for colonial affairs, Simon de Graaff, to the minister of waterworks, July 3, 1925; and report by the director of the Rijks Studiedienst voor Luchtvaart, E. B. Wolff, to the minister of waterworks, November 2, 1925: Archive of the Netherlands Ministry of Foreign Affairs, DEZ 151 *luchtvaart, bepalingen enz., "Vliegtuigen Nederland, controle op aanmaak enz."* dossier.

53. Report by the director of the Rijks Studiedienst voor Luchtvaart, E. B. Wolff, to the minister of waterworks, November 2, 1925: Archive of the Netherlands Ministry of Foreign Affairs, DEZ 151 *luchtvaart, bepalingen enz., "*Vliegtuigen Nederland, controle op aanmaak enz.*"* dossier.

54. Tony Fokker to Christine Döppler, August 20, 1921: Heitzman "Stuffinder" Aeronautical Collection, Deansboro, N.Y., Döppler album; Anthony Fokker, "De zeilvliegtuigwedstrijden in de Rhön," *Het Vliegveld 5,* no. 18 (August 27, 1921): 256–57.

55. *Le Matin,* November 17, 1921.

56. Report of W. Thorbecke, Netherlands chargé d'affaires in Paris, to minister of foreign affairs in The Hague, November 30, 1921: Archive of the Netherlands Ministry of Foreign Affairs, DEZ 151.

57. *Le Journal,* November 18, 1921.

58. Anthony Fokker, "Voorwoord," in W. M. van Neyenhoff and C. W. A. Oyens, *Zweefvliegen: Theorie en Praktijk* (Amsterdam: Meulenhoff, 1935), 7–8; C. W. A. Oyens, "Historische Tweezitters," *Avia* 18, no. 6 (1969): 288–89.

59. The possible detrimental effects of this delay were recognized by observers of Fokker's business conduct early on. In March 1921, RSL director E. B. Wolff pointed out the shortcomings of Fokker as a businessman before a governmental study commission on aeronautical development. "Beschouwingen van dr.ir. E. B. Wolff betreffende de aanbouw van vliegtuigen hier te lande," March 4, 1921: Archive of the Ministry of Waterworks, Commissie van Advies RSL, no. 58.

60. *Van Doorninck Report,* 9.

61. Ministère de la Guerre and Ministère des Traveaux Publics, *Les Aéronautiques Etrangères: Allemagne,* reports on the activities of "German" aircraft manufacturing companies for 1924 and 1926: Archives Nationales, Paris. In an

interview with the *New York Times,* Fokker presented a similar picture (May 30, 1926).

62. Stephan to Hegener, January 5, 1956. Typically, Stephan's arrival was not properly prepared by Fokker, and Stephan consequently had serious troubles establishing his newly given authority over the plant's general manager and export supervisor, Wilhelm Horter. Horter left discontented in September 1925. Horter to Hegener, October 1, 1925: Aeronautical Collection, Thijs Postma, Hegener Papers, correspondence.

63. Total turnover between 1925 and 1935 amounted to 22,800,000 guilders. All financial data here is from the *Van Doorninck Report,* 28–29.

64. Küpfer typescript, 90–91.

65. Van Vlissingen spoke of his rebuffed attempts to expand the capitalization of the Fokker company at the 155th meeting of the Board of Directors of KLM, December 17, 1935: KLM Board Papers, series R-5.

66. A full survey of research and development costs of Fokker airplanes has not survived, but in the second half of the 1920s, costs in guilders were as follows:

### Military Types

C.V (biplane/reconnaissance/light bomber, 1925): ƒ202,802.14
C.V-W (floatplane version of C.V, 1925–28): ƒ71,324.95
D.XIV (low-wing monoplane fighter, 1925): ƒ42,972.57
B.III (biplane/reconnaissance/flying boat, 1926): ƒ32,172.79
T.IV (monoplane/reconnaissance/bomber/floatplane, 1927): ƒ5,774.64
D.XVI (biplane fighter, 1929–32): ƒ8,427.52
C.VIII-W (monoplane/reconnaissance/bomber/floatplane, 1930): ƒ25,621.12
C.IX (sesquiplane/reconnaissance/light bomber, 1930): ƒ15,514.66
Total, military types: ƒ404,339.39

### Commercial Airliners

F.VII (single-engined, high-wing monoplane, 1925): ƒ57,338.22
F.VIII (twin-engined, high-wing monoplane, 1927): ƒ6,379.85
F.IX (three-engined, high-wing monoplane, 1929–30): ƒ29,144.88
F.XI (single-engined, high-wing monoplane, 1929): ƒ6,253.96
F.XII (three-engined, high-wing monoplane, 1930): ƒ10,956.72
F.XIV (single-engined, high-wing monoplane, 1929–32): ƒ92,138.06
Total, commercial airliners: ƒ202,221.69

Data were taken from a confidential report on the financial state of the Nederlandsche Vliegtuigenfabriek by the Netherlands state accountant, A. H. J. Schmitz, May 13, 1935: Archive of the Ministry of Finance, no. 1330-5/19.

67. Richard K. Smith, "Better: The Quest for Excellence," in Smithsonian Institution National Air and Space Museum, *Milestones of Aviation* (New York: Hugh Lauter Levin, 1989), 224–95. See also Wayne Biddle, *Barons of the Sky:*

*From Early Flight to Strategic Warfare: The Story of the American Aerospace Industry* (New York: Simon & Schuster, 1991), 147–60. On the rapid progress in American aeronautical design and engineering, see James R. Hansen, *Engineer in Charge: A History of the Langley Aeronautical Laboratory, 1917–1958* (Washington, D.C.: National Aeronautics and Space Administration, 1987).

68. See Henry L. Smith, *Airways: The History of Commercial Aviation in the United States,* reprint (Washington, D.C.: Smithsonian Institution Press, 1991), 158ff., 377–79.

69. In a letter to Hegener, Stephan later commented on the arrogant and condescending attitudes of both Fokker and Platz about the aerodynamic findings of university-trained engineers. Stephan to Hegener, July 17, 1957: Aeronautical Collection, Thijs Postma, Hegener Papers, correspondence. Marius Beeling, chief designer after Platz was forced to leave in 1931, painted a similar picture: Platz would draw a wing profile by hand, leaving it up to the younger engineers of the design bureau to come up with the aerodynamic characteristics of the wing. Beeling, "Herinneringen van oud-directeur en hoofdconstructeur Ir. Marius Beeling," *Avia* 18, no. 7 (1969): 328.

## CHAPTER 5. TO CAPTURE A WORLD MARKET

1. See Marc Dierikx, *Begrensde horizonten: Nederlandse burgerluchtvaartpolitiek in het interbellum* (Zwolle: Tjeenk Willink, 1988), 17–31; and "Struggle for Prominence: Clashing Dutch and British Interests on the Colonial Air Routes," *Journal of Contemporary History* 26 (1991): 333–51.

2. Minutes, twenty-fifth meeting of KLM's delegated board members, July 13, 1920: KLM Board Papers, series R-5.

3. KLM's corporate secretary, Hans Martin, to minister of waterworks, March 12, 1927: Archive of the Ministry of Transport and Waterworks, RLD, 14-I.

4. Henri Hegener, "Herinneringen aan Fokkervlieger Bernard de Waal," *Avia* 12, no. 4 (1963): 216–18.

5. G. J. Nijland, "Uit de jeugdjaren van de Nederlandsche Vliegtuigenfabriek," *Het Vliegveld* 12, no. 10 (1929): 365–66. Nijland was Fokker's brother-in-law.

6. See Marc Dierikx, "Hard Work Living off the Air: Albert Plesman's KLM in International Aviation, 1919–53," in William F. Trimble, ed., *Pioneers and Operators,* vol. 2 of *From Airships to Airbus* (Washington, D.C.: Smithsonian Institution Press, 1995), 221–44.

7. Plesman to the board of KLM, October 8, 1920: KLM Board Papers, series R-5.

8. Plesman to Fokker, December 30, 1920: KLM Board Papers, series R-5.

9. Minutes, forty-second meeting of delegated board members of KLM, June 7, 1921: KLM Board Papers, series R-5. In all, KLM would come to operate a fleet of fourteen F.III's.

10. Ibid.

11. Reinhold Platz to Alfred Weyl (author of *Fokker*), December 22, 1956: Aeronautical Collection, Peter M. Grosz, Weyl-Platz correspondence.

12. "Beschouwingen van dr.ir. E. B. Wolff betreffende de aanbouw van vliegtuigen hier te lande," March 4, 1921: Archive of the Ministry of Transport and Waterworks, Advisory Committee (Commissie van Advies) of the RSL, no. 58.

13. Customs records, July 1921, of aircraft imports into Holland between January 1, 1920, and June 1921: Archive of the Ministry of Transport and Waterworks, RLD, no. 625.

14. Deutsche Lufthansa, Firmenarchiv: Flugzeugzellen-Inventur per 31, December 1926 (courtesy R. E. G. Davies, NASM, blue binders).

15. Minutes, sixty-fifth meeting of KLM's delegated board members, March 13, 1923: KLM Board Papers, series R-5; minutes, sixth general meeting of KLM shareholders, April 17, 1923: KLM Board Papers, series R-501.

16. Fokker to KLM, May 25, 1923: KLM Board Papers, series R-5. The twin-engined plane was to be equipped with two 350 hp Rolls Royce Eagle engines, the three-engined plane with the 230 hp Siddeley Puma. In all likelihood, this is where Fokker's well-known series of successful three-engined aircraft originated.

17. Minutes, seventy-second meeting of KLM's delegated board members, November 2, 1923: KLM Board Papers, series R-5.

18. Sales contract between the Nederlandsche Vliegtuigenfabriek and KLM, December 10, 1923: KLM Board Papers, series R-5.

19. Plesman to Fentener van Vlissingen, February 1, 1924: KLM Board Papers, series R-5.

20. Confidential report, government accountant A. H. J. Schmitz to Van Doorninck Committee, May 13, 1935: Archive of the Ministry of Finance, 133-5(19).

21. "Lately, no matter how strange this may seem, we have become inclined to believe, that mister Fokker must be considered to be not quite normal. After all, the incoherent remarks, the unbusinesslike parlance, the foolish reproaches, that we have to hear from him in purely business meetings lately, have convinced us that mister Fokker cannot be held fully responsible for many of his words. We are not the only ones to reach this conclusion. Neither are we the only ones to complain about the strange conduct of mister Fokker. The same complaints can be heard at the Departments of War and of the Navy. Here too mister Fokker stands to lose all support." Plesman to KLM board member R. van Zinnicq Bergmann, April 23, 1924: KLM Board Papers, series R-5.

22. Plesman to KLM board member R. van Zinnicq Bergmann, April 23, 1924: KLM Board Papers, series R-5.

23. Marc Dierikx, *Bevlogen jaren: Nederlandse burgerluchtvaart tussen de wereldoorlogen* (Houten: Unieboek, 1986), 49–50, 86–89. The NVI went bankrupt in July 1926.

24. J. B. Scholte, flight report H-NADD for November 8 and 9, November 11, 1926: KLM Board Papers, series R-5.

25. The affair between Fokker and the NVI would not end with this cartel-like agreement. In 1930 S. del Monte, the main backer of the NVI (and signatory party in the agreement) secured a contract for license-production of six Curtiss fighter aircraft for the air force branch of the Dutch East Indies Army.

Fokker sued for 467,200 guilders in damages in a case that was settled by an arbitrator in 1934. Arbitrary decision in the case between Fokker/Nederlandsche Vliegtuigenfabriek and NVI/Del Monte, January 12, 1934: Archive of the Ministry of Finance, 1330-7(33).

26. Correspondence between Fokker and KLM, August 1925: KLM Board Papers, series A-9, Legal Affairs (Juridische Zaken) dossier.

27. Lesley Forden, *The Ford Air Tours, 1925–1931: A Complete Narrative and Pictorial History of the Seven National Air Tour Competitions for the Edsel B. Ford Reliability Trophy* (Alameda, Calif.: Nottingham Press, 1973), 4–5.

28. "Het eerste succesvolle Nederlandsche drie-motorige vliegtuig," *Fokker Bulletin* 2, no. 2 (November 1, 1925): 2. The plane was redesigned and completed in Amsterdam in a matter of weeks, after the engines arrived at the factory early in August. According to Stephan, Fokker's initial instructions were to place the additional engines in the wing roots, which Platz declined to do because it was necessary to get the plane shipped to the U.S. in time, and only minimal changes could be incorporated in the original F.VIIa wing layout. Bruno Stephan, "De Fokker F.VII Series," *Avia-Vliegwereld* 5, no. 2 (1956): 33. See also Hegener, *Fokker,* 55–58.

29. Stephan to Hegener, August 23, 1965: Aeronautical Collection, Thijs Postma, Hegener Papers, correspondence.

30. "De drie-motorige Fokker in de 'Commercial Airplane Reliability Tour' om de 'Ford Trophy,'" *Fokker Bulletin* 2, no. 2 (November 1, 1925): 3–4; see also Henry M. Holden, *The Fabulous Tri-Motors* (Blue Ridge Summit, Penn.: TAB Books, 1992), 45–47.

31. B. van der Klaauw, "Friendship Voorlopers," pt. 2, *Avia- Vliegwereld* 9, no. 15 (1960): 414–15; Hegener, *Fokker,* 189–90.

32. The F.XII cost Fokker 10,957 guilders in research and development. Confidential report on the financial state of the Nederlandsche Vliegtuigenfabriek by the Netherlands State Accountant, A. H. J. Schmitz, May 13, 1935: Archive of the Ministry of Finance, no. 1330-5(19). Limited license production of a military variant of the F.IX was conducted by Avia in Czechoslovakia. Hegener, *Fokker,* 191.

33. Photographs of KNVvL gatherings and the arrival of the first London-Amsterdam flight, *Het Vliegveld,* May 1920.

34. The extraordinary frequency of his travels is evident in visa and border stamps in a number of his Dutch passports, which have survived in the Aviodome Museum Collection at Schiphol Airport, Amsterdam.

35. Geert J. Nijland, "Bij Tonny Fokker's dood," *Het Vliegveld* 24, no. 1 (1940): 20; W. P. van den Abeelen (captain in the Air Corps and a prominent aviation journalist on the side), "In Memoriam Anthony H. G. Fokker," *Het Vliegveld* 24, no. 1 (1940): 4; memories of Anthony Fokker, written down for Hegener by Fokker's sister, Toos, undated (ca. 1951): Aeronautical Collection, Thijs Postma, Hegener Papers, dossier of Fokker material. See also the interview by the author with Fokker's nephew, Ton Nijland, February 19, 1987.

36. Döppler photo albums and correspondence: Heitzman "Stuffinder" Aeronautical Collection.

37. Wim van Neijenhoff's personal recollections of Anthony Fokker, written

down for Hegener on March 1, 1940: Aeronautical Collection, Thijs Postma, Hegener Papers, correspondence; memories of Toos Fokker, written down for Hegener, undated (ca. 1951): Aeronautical Collection, Thijs Postma, Hegener Papers, dossier of Fokker material.

38. Henri Hegener, "Mijn herinneringen aan Anthony Fokker: De groote luchtvaartpionier zooals hij was als mensch en als vriend," *De Week in Beeld,* March 1940, 9–11 (signed "luchtvaarder": aviator).

39. Reminiscences of Fokker's private secretary, Cis van Rhijn, in a letter to Hegener, April 16, 1957: Aeronautical Collection, Thijs Postma, Hegener Papers, correspondence; scribbled menu card from a sea trip to Sweden in 1923 ("Enjoying the party on board H M 'Leviathan' with the two nicest Schwedisch girls on a high sea and under a good dinner," signed by Tony Fokker, Iris Fagerström, Ruth Requello, and Henri Hegener); postcard, Tony Fokker to his mother ("Warmest regards from the land of the snow, the sun, the sport, the flirt, and the luxury."): Aviodome Archive, Anthony Fokker dossier, correspondence. Recollections of Fokker by his contemporaries tend to stress the same behavior.

40. In February 1940, Tetta came back to Holland to witness the interring of Fokker's ashes at the Westerveld cemetery near Haarlem. Life's burdens showed on her then, and to Tony's mother, she admitted being near the end of her financial resources. Anna Fokker-Diemont to Christine Döppler, June 21, 1941: Heitzman "Stuffinder" Aeronautical Collection, Döppler album.

41. Notes, Toos Fokker on her brother, written down for Hegener, undated (ca. 1951): Aeronautical Collection, Thijs Postma, Hegener Papers, dossier of Fokker material.

42. Nijland, "Bij Tonny Fokker's dood," 20.

43. Photographs of Oberalpina: Aeronautical Collection, Thijs Postma, Hegener Papers, dossier of Fokker material; interview by the author with Ton Nijland, February 19, 1987.

44. Memories written down by Toos Fokker for Hegener, undated (ca. 1951): Aeronautical Collection, Thijs Postma, Hegener Papers, dossier of Fokker material.

45. Cis van Rhijn to Hegener, April 16, 1957: Aeronautical Collection, Thijs Postma, Hegener Papers, correspondence.

46. Interview by the author with Ton Nijland, February 19, 1987.

47. Minutes, seventy-fifth meeting of KLM's delegated board members, February 15, 1924: KLM Board Papers, series R-5.

48. Official report by Smirnoff of the accident with the H-NABH on the Goodwin Sands, with an extract from the ship's log of the *Primo* dated October 19, 1923: Archive of the Netherlands Ministry of Foreign Affairs, DEZ 151 *Luchtvaart, bepalingen enz.* See also Iwan W. Smirnoff, *Smirnoff vertelt* (Amsterdam: Blitz, 1938), 187–96.

49. Hegener, *Fokker,* 43.

50. Gen. Cornelis Snijders in a speech to the Committee Netherlands—Indies Flight, March 26, 1927, in *Mededeelingenblad Comité Vliegtocht Nederland-Indië* 4 (April 1927): 64 (emphasis added). Despite careful preparations of cotton ear plugs, even a famous flyer like Kingsford-Smith attested to this

problem: "Then came a moment when they sought to take us away from the machine. We protested with vigor, as the throng that surged about us looked sufficiently excited to pluck off a propeller or some such trifle as a souvenir. Various people in authority bawled into our ears, and finally got the message through the waning engine thunder in our heads that the Police and the Defence Force were mounting guard over the plane, and that we could go and sleep in contentment." Charles E. Kingsford-Smith and Charles T. P. Ulm, *The Flight of the Southern Cross* (New York: McBride, 1929), 199.

51. Betty Stettinius Trippe, *The Diary and Letters of Betty Stettinius Trippe, 1925–1968* (New York, privately published, 1982), 31.

52. The flight is recounted in detail in A. N. J. Thomassen à Thuessinck van der Hoop, *Door de lucht naar Indië* (Amsterdam, 1925).

53. Accountant's report, W. Kreukniet and R. A. Dijker Accountancy, on the Comité Vliegtocht Nederland-Indië, April 30, 1925: Wm. Teuben "Teubarium," Aeronautical Collection, Voorburg, Netherlands.

54. George Hubert Wilkins, *Flying the Arctic* (New York: Putnam, 1928), 13–15.

55. Ibid., 29–48.

56. Richard E. Byrd, *Skyward: Man's Mastery of the Air as Shown by the Brilliant Flights of America's Leading Air Explorer: His Life, His Thrilling Adventures, His North Pole and Transatlantic Flights, together with His Plans for Conquering the Antarctic by Air* (New York: Halcyon House, 1928), 166–206. Byrd's claim to have actually reached the pole was later publicly disclaimed, upon recalculation of weather conditions, average cruising speeds, etc. by the Swedish professor of meteorology at the University of Uppsala, Gösta H. Liljequist, who put the northernmost point reached by the *Josephine Ford* at 88° 17.5'. See Richard Montague, *Oceans, Poles, and Airmen: The First Flights over Wide Waters and Desolate Ice* (New York: Random House, 1971), 281–88 (appendix A, reprinted from Liljequist's original article in *Interavia: World Review of Aviation and Aeronautics,* May 1960). Similar statements by Byrd's Norwegian copilot, Bernt Balchen, in the manuscript version of his autobiography, *Come North with Me* (New York: Dutton, 1958), claiming that Byrd had manufactured navigational data and that they turned back several hundred miles from the pole, had to be changed before distribution of a revised edition of the book under threat of a libel suit from the Byrd family. Montague, *Oceans,* 289–300 (where Montague compares the two versions of Balchen's text).

57. Byrd, *Skyward,* 234–38 (with photographs taken after the crash); Fokker and Gould, *Flying Dutchman,* 255–59.

58. Byrd, *Skyward,* 239: "That crash-fate had engineered me into a tight place. My stock was down to zero."

59. Fokker and Gould, *Flying Dutchman,* 262. Balchen elaborates on the incident in *Come North,* 100.

60. A detailed account of the flight based on an interview with Noville appeared in the Dutch aeronautical journal *Het Vliegveld* 11, no. 7 (1927): 213–18. See also Byrd, *Skyward,* 222–78; and Fokker and Gould, *Flying Dutchman,* 255–65.

61. An authentic account of the flight by Scholte appeared as "Amsterdam-Batavia en terug" in Pieter Guilonard and Gijs Spit, *De weg door de lucht: Het moderne snelverkeer* (Amsterdam, 1927), 314–18.

62. G. A. Koppen, *Holland-Indië-Holland, in storm en zonnebrand* (Amsterdam: Telegraaf, 1927).

63. Kingsford-Smith and Ulm, *Flight*, 14–30. See also Gregory C. Kohn, "The 'Southern Cross' Complex," *American Aviation Historical Society Journal* 4, no. 3 (1959): 195–206.

64. The flight is extensively described in Kingsford-Smith and Ulm, *Flight*.

65. Beeling, "Herinneringen van oud-directeur en hoofdconstructeur Ir. Marius Beeling," *Avia* 18, no. 7 (July 1969): 329.

## CHAPTER 6. THE AMERICAN ENCHANTMENT

1. James C. Fahey, *U.S. Army Aircraft (Heavier than Air), 1908–1946* (New York: Ships and Aircraft, 1946), 19.

2. Peter M. Bowers, "The American Fokkers," pt. 1, *AAHS Journal* 11, no. 2 (1966): 79–92.

3. Interview with Anthony Fokker in *The World*, November 9, 1919.

4. *New York Times*, November 11, 1920.

5. The exact date the company was founded remains unclear. The Netherlands Aircraft Manufacturing Company first appears in the Trade Index of the 1922 *Aircraft Year Book* (250), which summarizes the state of aeronautical affairs in the U.S. in 1921.

6. By one estimate, 75 percent of all radios and 60 percent of all cars and furniture were paid for in installments. American Social History Project, *Who Built America? Working People and the Nation's Economy, Politics, Culture and Society*, vol. 2, *From the Gilded Age to the Present* (New York: Pantheon Books, 1992), 275.

7. American Social History Project, *Who Built America?* 269–90. See also Sean Cashman, *America in the Twenties and Thirties: The Olympian Age of Franklin Delano Roosevelt* (New York: New York University Press, 1989), 5–65.

8. Helen Kay Schunck, "Working with a Genius," *Aeronautics*, August 1929, 76.

9. Fokker to Christine Döppler, undated (December 1920), Heitzman "Stuffinder" Aeronautical Collection, Döppler album.

10. Bowers, "American Fokkers," pt. 1, 85–89.

11. Küpfer typescript, 76.

12. Burke Davis, *The Billy Mitchell Affair* (New York: Random House, 1967), 133.

13. Bowers, "American Fokkers," pt. 1, 86–91. Bowers's account is based on Mitchell's report of his European mission. *Report of Inspection Trip to France, Italy, Germany, Holland, and England, Made during the Winter of 1921–1922*, Air Service Information Circular, vol. 4, no. 391, November 1, 1922; Davis, *Billy Mitchell Affair*, 139–40.

14. Mitchell to Gen. Mason Patrick, quoted in Davis, *Billy Mitchell Affair,* 140. Although Davis puts Mitchell's words in quotation marks, his system of annotation does not make it possible to ascertain the archival source.

15. Postcard, Tony Fokker to his mother, May 13, 1922: Aviodome Archive, Letters and Postcards Anthony Fokker dossier.

16. *New York Times,* July 23, 1922.

17. Hubrecht to minister of foreign affairs, July 26, 1922: Archive of the Netherlands Ministry of Foreign Affairs, DEZ 151.

18. Nick A. Komons, *Bonfires to Beacons: Federal Aviation Policy under the Air Commerce Act, 1926–1938* (Washington, D.C.: Smithsonian Institution Press, 1989), 7–33. For an in-depth analysis of the daredevil element and the public image of aviation, see Dominick Pisano's forthcoming article, "The Greatest Show Not on Earth: The Confrontation between Utility and Entertainment in Aviation" (Washington, D.C.: National Air and Space Museum).

19. *New York Times,* July 23, 1922, and August 4, 1922.

20. Noorduyn to Arcier, May 22, 1924: National Air and Space Museum, Arcier Collection, box 1, correspondence; H. A. Seyffardt, "De ontwikkeling van het Fokker-concern in Amerika," *Het Vliegveld* 13, no. 2 (1929): 58–61; Hegener, *Fokker,* 68; Peter M. Bowers, "The American Fokkers," pt. 2, *AAHS Journal* 11, no. 4 (1966): 253–55. Arcier left Fokker's employ on February 1, 1928, to join the General Airplane Corporation of Buffalo, New York.

21. C. G. Grey, ed., *Jane's All the World's Aircraft* (London: Sampson, Low, Marston, 1920), 299; and 1922 edition, 196b; G. Allison Long Jr., "The Biography of A. Francis Arcier (1890–1969)," *AAHS Journal* 26, no. 2 (1981): 126–42.

22. Küpfer typescript, 76.

23. Seyffardt, "Ontwikkeling," 58. The turnover for the first six months was $72,184. See also John Sloan, "Fokker in the United States," pt. 1, *Aeronautica: Official Publication for Historical Associates of the Institute of Aeronautical Sciences* 7, no. 1 (1955): 3–5.

24. Stephan to Hegener, August 30, 1958: Aeronautical Collection, Thijs Postma, Hegener Papers, correspondence.

25. Interview, *New York Times,* November 8, 1924.

26. Bowers, "American Fokkers," pt. 2, 256–58.

27. The Air Mail Act also drew attention to the wider legal issue of regulating civil aviation. Though the first proposals dated from March 1916, the Air Commerce Act was only signed into law on May 20, 1926. Komons, *Bonfires,* 66–124. See also Carl Solberg, *Conquest of the Skies: A History of Commercial Aviation in America* (Boston: Little, Brown, 1979), 63–65.

28. Smith, *Airways,* 94–116.

29. Plesman to KLM board member R. van Zinnicq Bergmann, reporting on a discussion with Fokker, April 23, 1924: KLM Board Papers, series R-5.

30. "Fokker–Kansas City Company. Fokker Plans to Become Resident of the USA and Build Here," *Aviation,* August 31, 1925. See also Seekatz to A. Zimmermann (Fokker's lawyer in Berlin), May 3, 1933: Fokker Archive, V/Alg. Secr., no. 67.

31. Tony Fokker to Christine Döppler, undated (June 1925): Heitzman "Stuffinder" Aeronautical Collection, Döppler album.

32. "The Fokker–Kansas City Company." See also Hegener, *Fokker,* 70.

33. Forden, *Ford Air Tours,* 18–19.

34. Bowers, "American Fokkers," pt. 2, 258–60.

35. Ibid.

36. Matthew Josephson, *Empire of the Air: Juan Trippe and the Struggle for World Airways* (New York: Harcourt, Brace, 1944), 24, 28–29; Wesley P. Newton, *The Perilous Sky: Evolution of United States Aviation Diplomacy toward Latin America, 1919–1931* (Coral Gables, Fla.: University of Miami Press, 1978), 110; Robert Daley, *An American Saga: Juan Trippe and His Pan Am Empire* (New York: Random House, 1980), 15–19.

37. Photo of Tony Fokker taken in the late 1920s inscribed "Der Flieger Fokker am Arbeitstisch." Hegener, *Fokker,* 2.

38. Sales brochure, Fokker Aircraft Corporation of America, undated (1929): Aviodome Archive, Fokker Collection.

39. The reader interested in this confusing matter is best advised to read Peter Bowers's five-part series, "The American Fokkers," which appeared between 1966 and 1971 in the *AAHS Journal.*

40. In 1929, the F-14 combined a mail plane and a small airliner—another example of the peculiar approach to aircraft design at Fokker Aircraft Corporation. Looking like a reverse-engineered F.V driven by a modern engine, the plane appeared to combine a shortened F.VIIa wing, tail section, and undercarriage with the type of fuselage layout and parasol-wing attachment tried in 1922. Fourteen were built.

41. Elsbeth E. Freudenthal, *The Aviation Business: From Kitty Hawk to Wall Street* (New York: Vanguard Press, 1940), 98–100; Grover Loening, *Our Wings Grow Faster* (Garden City, N.Y.: Doubleday, 1935), 169–94; Claude E. Puffer, *Air Transportation* (Philadelphia: Blakiston, 1941), 514.

42. *New York Times,* December 19, 1927, and May 25, 1928. See also Bayard Young, "Ohio Valley Weekend 50th: First W. Va. Commercial Plane Flew in 1928," *Intelligencer* (Wheeling, W. Va.), December 16, 1978.

43. *New York Times,* May 25, 1928, and October 25, 1928. See also Young, "Ohio Valley Weekend."

44. Freudenthal, *Aviation Business,* 88–92; Puffer, *Air Transportation,* 515. In his famous book *Airways,* Smith calculates that to justify the level of investments in 1929, the aviation sector should have sold about 25,000 private and commercial aircraft, whereas actual sales amounted to 3,500. Smith, *Airways,* 131.

45. Joseph J. Corn, *The Winged Gospel: America's Romance with Aviation, 1900–1950* (New York: Oxford University Press, 1983), 91–111.

46. R. E. G. Davies, *Airlines of the United States since 1914* (London: Putnam, 1972), 43–45.

47. Davies, *Airlines,* 58–69.

48. *New York Times,* October 25, 1928.

49. Ibid., May 17, 1929.

50. Wall Street indexes of "Over the Counter Quotations for Unlisted Securities," *New York Times,* October 1928.

## CHAPTER 7. RUNNING ON EMPTY

1. Interviews with Tony Fokker in the *New York Times,* May 30, 1926, and December 16, 1928: "The reason for my decision two years ago to become an American citizen [was] in order that I might devote the remainder of my career to the advancement of the aeronautical industry here." See also Schunck, "Working with a Genius."

2. Tony Fokker to Christine Döppler, undated (summer 1925): Heitzman "Stuffinder" Aeronautical Collection, Döppler album. Döppler had originally come into contact with Fokker because of her position with the Von Morgens as a lady companion for Tetta, whose mother had died young.

3. Schunck, "Working with a Genius," 74.

4. Ibid., 76.

5. Doree Smedley and Hollister Noble, "Profiles: Flying Dutchman," *New Yorker,* February 7, 1931, 20–24.

6. Balchen, *Come North,* 68–71.

7. Smedley and Noble, "Profiles," 20–24.

8. Interview with Tony Fokker in the *New York Times,* December 12, 1925.

9. Fokker's presentation before the American Society of Mechanical Engineers (Buffalo, N.Y.), *New York Times,* April 26, 1927; Fokker's presentation before the Netherlands Chamber of Commerce in New York, *New York Times,* October 19, 1927; interview with Tony Fokker in *Chicago Herald & Examiner,* August 11, 1929. For a wider context, see Komons, *Bonfires,* 244–48.

10. Only one of each was built. The two-seat, open-cockpit *Skeeter* (model 8) appeared late in 1926. In 1928 work on the model 11 (also known as H-51) was contracted out to the Hall Aluminium Aircraft Corporation, a small company that had experience in all-metal construction. The model 11 was a rather innovative small plane incorporating a three-seat enclosed cockpit and wings that could be folded to fit the plane into a car garage. A parallel design for a two-seater, also tested in 1929, was known as model 13. Peter F. A. van de Noort, *Fokkers "Roaring Twenties": De vliegtuigen van de Amerikaanse Fokker-fabrieken* (Sassenheim: Rebo Produkties, 1988), 43–47; Peter M. Bowers, "The American Fokkers," pt. 3, *AAHS Journal* 12, no. 3 (1967): 180; and pt. 4, *AAHS Journal* 16, no. 3 (1971): 190–92.

11. Interview with Tony Fokker in the *New York Times,* May 30, 1926.

12. W. T. Whalen, vice chairman of the board, in a Fox Movietone Newsreel, March 1930, upon christening of the first F-32: National Air and Space Museum film archive.

13. Interview with Tony Fokker in the *New York Times,* December 2, 1928.

14. The exact date of the marriage is still unknown, but the state of Tony Fokker's clothes in various surviving photographs from that time suggests a

sudden increase of feminine influence on his attire in July 1927. A photograph showing Fokker and U.S. Army Air Service lieutenants Maitland and Hegenberger shortly after Hegenberger and Maitland's pioneering flight from Oakland to Hawaii on June 28–29 (Hegenberger is holding up a small case with a commemorative medal) shows Fokker still in his usual rumpled vestments. Hegener, *Fokker,* 68. Fokker also took out life insurance for a surprisingly low $50,000 with the United States Life Insurance Company in early August, perhaps for Violet's benefit. *New York Times,* August 4, 1927. Newspaper reports after Violet's death in February 1929 stated they had been married about a year and a half. *New York Times,* February 10, 1929.

15. *New York Times,* February 9, 1929.

16. Joan Jacobs Brumberg, "Chlorotic Girls, 1870–1920: A Historical Perspective on Female Adolescence," in Judith Walzer Leavitt, ed., *Women and Health in America: Historical Readings* (Madison: University of Wisconsin Press, 1984), 186–95. Brumberg, page 190, refers to B. Sicherman, "The Uses of Diagnosis: Doctors, Patients and Neurasthenia," *Journal of the History of Medicine and Allied Sciences* 32 (1977): 33–54. I am indebted to Dixie Dysart of Auburn University for taking me through the medical history of neurasthenia.

17. Clayton L. Thomas, ed., *Taber's Encyclopedic Medical Dictionary* (Philadelphia: F. A. Davis, 1989), 1200.

18. Reconstructed from eyewitness accounts given to reporters, *New York Times,* February 9 and 10, 1929.

19. The report read "almost distracted." *New York Times,* February 9, 1929.

20. *New York Times,* February 10, 1929.

21. Peter M. Grosz, "Propwash: An Open Letter to the Editor," *Windsock International* 3, no. 3 (1987): 2. Grosz was Gould's neighbor at one time after the latter retired as the *New York Evening Post*'s aviation journalist. See also Bruce Gould, "Herinneringen aan Tony Fokker," *Het Vliegveld* 25, no. 1 (1941): 14. The original edition of the autobiography was published in New York in April 1931.

22. Fokker and Gould, *Vliegende Hollander,* 111.

23. Biddle, *Barons of the Sky,* 178.

24. For a recent and comprehensive study based on extensive archival research on the fortunes of the U.S. aircraft industry between 1918 and 1935, see Biddle, *Barons of the Sky,* 156–93. See also Freudenthal, *Aviation Business,* 97–99.

25. Alfred P. Sloan Jr., *My Years with General Motors* (Garden City, N.Y.: Doubleday, 1963), 362–63.

26. *New York Times,* New York Stock Exchange stock price fluctuations and news items, April 1–May 19, 1929. Nevertheless, Universal was taken over by the Aviation Corporation. Baker, Simonds & Co., *The Aviation Industry* (Detroit, Baker, Simonds, 1929), 20–22.

27. Sloan, *My Years with General Motors,* 364; *New York Times,* New York Stock Exchange share prices, April 29, 1929, to May 19, 1929; articles in the *New York Times* financial section on the GM takeover bid, May 17 and July 24, 1929.

28. *New York Times,* July 24 and 27, 1929. See also Edward V. Rickenbacker, *Rickenbacker: An Autobiography* (Englewood Cliffs, N.J.: Prentice-Hall, 1967), 178–79; Peter M. Bowers, "The American Fokkers," pt. 4, *AAHS Journal* 16, no. 3 (1971): 189.

29. *New York Times,* June 23, 1929.

30. Ibid., October 23, 1929.

31. Ibid.

32. Ibid., October 24, 1929.

33. Ibid., May 21, 1930.

34. Komons, *Bonfires,* 219.

35. Jim King to Los Angeles office of WAE, November 23, 1929, cited in Robert H. Schleppler, "The Fokker F-32," *AAHS Journal* 11, no. 2 (1966): 110–17 (quotation on 111).

36. Schleppler, "Fokker F-32," 112. See also Rickenbacker, *Rickenbacker,* 179.

37. René de Leeuw, ed., *Fokker Verkeersvliegtuigen: Van de F.I uit 1918 tot en met de Fokker 100 van nu* (Amsterdam: NV Koninklijke Nederlandse Vliegtuigenfabriek Fokker, 1989), 132.

38. Sales brochure, Fokker Aircraft Corporation of America, effective April 7, 1930: Aviodome Archive, Fokker Collection.

39. Davies, *Airlines,* 593–601.

40. Sloan, *My Years with General Motors,* 364.

41. *New York Times,* May 26, 1930, and June 19, 1930.

42. Young, "Ohio Valley Weekend."

43. *New York Times,* April 7, 1930.

44. Ibid., May 26, 1930. On June 18, 1930 GM turned over the Fokker stock to General Aviation Corporation, and on June 30, the Fokker manufacturing rights came to reside with the General Aviation Manufacturing Corporation, although the Fokker name was not actually dropped until August 19, 1931. Information supplied to the author by North American/Rockwell, December 19, 1984.

45. *New York Times,* June 16, 1930.

46. Information supplied to the author by North American/Rockwell, December 19, 1984.

47. Biddle, *Barons of the Sky,* 96–106, 240. See also Kenneth Munson, *Airliners between the Wars, 1919–1939* (London: Blandford Press, 1972), 158.

48. *New York Times,* July 24, 1930.

49. Davies, *Airlines,* 89–93.

50. Bowers, "American Fokkers," pt. 4, *AAHS Journal,* 16, no. 3 (1971): 189–95. See also Peter M. Bowers, "The American Fokkers," pt. 5, *AAHS Journal* 16, no. 4 (1971): 252–62. A series of twelve YO-27 experimental twin-engined bombers was manufactured at General Aviation's new production center at the former Curtiss-Caproni plant in Dundalk, just outside Baltimore, Maryland.

51. *New York Times,* March 22, 1931.

52. Testimony of C. H. McCracken, April 1, 1931: National Air and Space Museum, Rockne Crash file.

53. For a detailed evaluation of the accident and subsequent events, see Dominick A. Pisano, "The Crash that Killed Knute Rockne," *Air & Space Smithsonian* 6, no. 5 (1991/1992): 88–93.

54. Pisano, "Crash," 88.

55. Leonard Jurden (the bureau's supervising inspector at the scene) to G. C. Budwig (director of air regulation at the Department of Commerce), April 8, 1931: National Air and Space Museum, Aeronautics Department, Rockne Crash file.

56. Komons, *Bonfires,* 185. Komons quotes from a 1971 interview with Young in which he claimed the bureau had "missed the boat by one day" in issuing a grounding order for the F-10.

57. Tony Fokker to the Executive Committee of Trans-Continental and Western Air Express, April 6, 1931: National Air and Space Museum, Aeronautics Department, Rockne Crash file.

58. Ibid.

59. Jurden to the chief of the Inspection Service of the Aeronautics Branch, April 7, 1931: National Air and Space Museum, Aeronautics Department, Rockne Crash file.

60. Young, "Ohio Valley Weekend."

61. Komons, *Bonfires,* 185–88; Pisano, "Crash," 93.

62. Hegener, *Fokker,* 76; Thijs Postma, *Fokker: Aircraft Builders to the World* (London: Jane's Publishing, 1980), 87; Komons, *Bonfires,* 189. More recently, see Jacob A. Vander Meulen, *The Politics of Aircraft: Building an American Military Industry* (Lawrence: University of Kansas Press, 1991), 95. There are many other examples.

63. *New York Times,* July 11 and 13 and August 22, 1931; Carter Tiffany, "A Personal Memory of Tony Fokker," undated (ca. 1940): Aeronautical Collection, Thijs Postma, Hegener Papers, correspondence.

64. *New York Times,* March 24 and April 25, 1933; information supplied to the author by North American/Rockwell, December 19, 1984.


## CHAPTER 8. WEATHERING THE STORM

1. Eric Schatzberg, "Ideology and Technical Choice: The Decline of the Wooden Airplane in the United States, 1920–1945," *Technology & Culture* 35, no. 1 (1994): 34–69 (quoted on 52).

2. During the Second World War, the example of the wood-built British de Havilland Mosquito showed that for medium-size aircraft, the real issue at stake was not necessarily the choice between wood and metal, but the adoption of advanced aerodynamics in design and modern structural practices that used the aircraft's skin to distribute stress loads experienced in flight evenly over the airframe.

3. Hansen, *Engineer,* 123–32.

4. Minutes, ninth meeting of the Van Doorninck Committee, November 8, 1934: Archive of the Ministry of Finance, dossier 1330-4(16).

5. Second report of the Van Doorninck Committee, appendix 2, June 21, 1935: Archive of the Netherlands Ministry of Waterworks, RLD, 14-III.

6. *Wood or Metal? Holz oder Metall? Bois ou Metal?* Trilingual brochure published by the NV Nederlandsche Vliegtuigenfabriek Fokker, dated May 1938: Fokker Archive (Verhoef), series VO/PT, box 3.

7. Plesman to the board of KLM, December 8, 1930: KLM Board Papers, series R-5.

8. Minutes, Fentener van Vlissingen in the 155th board meeting of KLM, December 17, 1935: KLM Board Papers, series R-5.

9. Minutes, 100th board meeting of KLM, September 27, 1932. See also Tony Fokker to Plesman, October 13, 1932: KLM Board Papers, series R-5.

10. Hansen, *Engineer,* 148–58. For the wider context of the technical development of the new generation of transport aircraft, see Peter W. Brooks, *The Modern Airliner: Its Origins and Development* (London: Putnam, 1961), 67–86. See also John B. Rae, *Climb to Greatness: The American Aircraft Industry, 1920–1960* (Cambridge: MIT Press, 1968), 61–71.

11. Douglas J. Ingells, *The McDonnell-Douglas Story* (Fallbrook, Calif.: Aero Publishers, 1979), 46. Ingells quotes the Douglas company as his source.

12. For comparison, the Fokker F-10A sold for $67,500 in the spring of 1929. Sales brochure, Fokker Aircraft Corporation of America, April 1930: Aviodome Archive, Fokker Collection. Brooks's often-quoted assertion that the DC-2 was an expensive plane was incorrect: Lockheed's smaller (eight-to-ten-seat) L-10 Electra sold for $50,000. Brooks, *Modern Airliner,* 83. See also Rae, *Climb to Greatness,* 66 and 71; and Wilbur H. Morrison, *Donald W. Douglas: A Heart with Wings* (Ames: Iowa State University Press, 1991), 80.

13. Stephan to Netherlands minister of foreign affairs, Beelaerts van Blokland, October 23, 1931: Archive of the Netherlands Ministry of Foreign Affairs, DEZ 70B Romania.

14. Correspondence on file with the Netherlands Ministry of Foreign Affairs, DEZ 70B Romania.

15. Oudendijk to Netherlands minister of foreign affairs, October 1931: correspondence on file with the Netherlands Ministry of Foreign Affairs, Archive of the Netherlands Legation in Peking, series V, 1931–41, dossier P.IV (box 318) and dossier DEZ III, 336.

16. Correspondence on file with the Netherlands Ministry of Foreign Affairs: Archive of the Netherlands Legation in Peking, series V, 1931–41, dossier P.IV (box 318) and dossier DEZ III, 336.

17. Minutes, 100th board meeting of KLM, September 1932; Anthony Fokker to Plesman, October 13, 1932: KLM Board Papers, series R-5.

18. Oudendijk, commander of the Air Corps, before the Van Doorninck Committee, January 24, 1935: Archive of the Ministry of Finance, dossier 1330-4(16).

19. Minutes, sixteenth meeting of the Van Doorninck Committee, May 21, 1935: Archive of the Ministry of Finance, dossier 1330-4(16); financial data from the Küpfer typescript, 96.

20. Anthony Fokker to Van Doorninck Committee, September 6, 1934:

Archive of the Ministry of Finance, dossier 1330-6(26). The depth of the crisis in 1932 was also reflected in the number of Fokker test flights at Schiphol Airport, which fell from 402 in 1930 to a mere 62 in 1932: Annual Reports, Gemeente Handels Inrichtingen Amsterdam, Schiphol Airport, 1930 and 1932.

21. See Marc Dierikx and Peter Lyth, "The Development of the European Scheduled Air Transport Network, 1920–1970: An Explanatory Model," in Albert Carreras, Andrea Giuntini, and Michèle Merger, eds., *European Networks, Nineteenth–Twentieth Centuries: New Approaches to the Formation of a Transnational Transport and Communications System,* Proceedings Eleventh International Economic History Congress, vol. B.8 (Milan: Università Bocconi, 1994), 73–93.

22. Van Doorninck Report, Annex 2, June 21, 1935: Archive of the Ministry of Waterworks, RLD, 14-III.

23. Lambert H. Slotemaker (head of KLM's legal bureau) to minister of waterworks, December 1, 1933: Archive of the Ministry of Waterworks, RLD, 14-III.

24. Accountant's report on the Fokker factory by Tekelenburg & Van Til Accountancy, March 7, 1935: Archive of the Ministry of Finance, dossier 1330-6(25).

25. Tony Fokker to Plesman, October 13, 1932: KLM Board Papers, series R-5.

26. Confidential report by the government's accountant, A. H. J. Schmitz, on the past financial performance of the Fokker Company for the Van Doorninck Committee, May 13, 1935: Archive of the Ministry of Finance, 1330-5(19).

27. KLM maintenance division, report no. 10 on the F.XXXVI, September 13, 1934; confidential survey of the problems with PH-AJA (as the aircraft was registered) from March 16, 1935, to about October 1, 1935, November 20, 1935: KLM Board Papers, series R-5.

28. Minutes, KLM board meeting, November 28, 1933: KLM Board Papers, series R-5.

29. Albert Plesman Jr., *Albert Plesman, mijn vader* (The Hague: Nijgh & Van Ditmar, 1977), 41–42.

30. Plesman to minister of waterworks, November 29, 1933: Archive of the Ministry of Transport and Waterworks, RLD, 14-II.

31. Agreement between Donald Douglas; his secretary, I. L. MacMahon; and Anthony Fokker, January 15, 1934: Fokker Archive, V/Alg. Secr., no. 15.

32. Ibid., arts. 1, 18, and 19.

33. Anthony Fokker to Van Doorninck, August 16, 1934: Archive of the Ministry of Finance, dossier 1330-6(26).

34. Minutes, second meeting of the Van Doorninck Committee, February 15, 1934: Archive of the Ministry of Finance, dossier 1330-4(16).

35. Plesman to the members of the board of KLM, June 21, 1934; minutes, 130th and 131st meetings of the board of KLM, June 26 and July 28, 1934: KLM Board Papers, series R-5.

36. Fokker to Van Doorninck, August 16, 1934: Archive of the Ministry of Finance, dossier 1330-7(33).

37. Plesman to Fokker, August 23, 1934: KLM Board Papers, series R-5.

38. Minutes, second meeting of the Van Doorninck Committee, February 15, 1934: Archive of the Ministry of Finance, dossier 1330-4(16). The argument was repeated time and again, more diplomatically, by a number of people from KLM's higher echelons in the spring of 1934.

39. Minutes, 134st board meeting of KLM, October 30, 1934: KLM Board Papers, series R-5.

40. Dierikx, *Begrensde horizonten,* 116–32. See also Dierikx, "Struggle for Prominence: Clashing Dutch and British Interests on the Colonial Air Routes, 1918–1942," *Journal of Contemporary History* 26 (1991): 333–51.

41. Koene D. Parmentier, *In drie dagen naar Australië: Met de Uiver in de Melbourne Race* (Amsterdam: Scheltens & Giltay, 1935).

42. Minutes, KLM board meeting, October 30, 1934: KLM Board Papers, series R-5.

43. Contract, KLM-Fokker on the DC-2 sale, August 15 and December 31, 1935: KLM Board Papers, series A-950; overview of price negotiations between KLM and Fokker regarding the DC-2, March 14, 1939: Archive of the Ministry of Transport and Waterworks, RLD, 14-V.

44. Official statement of Donald Wills Douglas, May 23, 1935: KLM Board Papers, series A-9/A-950, R-506-2.

45. Anthony Fokker to Donald W. Douglas, June 5, 1935: Fokker Archive, V/Alg. Secr., no. 15. The assertion in Morrison's *Donald W. Douglas* that "adding to his [Douglas's] joy of the trip was meeting with his old friend Tony Fokker" (84) could hardly be farther from the truth.

46. Interview with Anthony Fokker, *De Maasbode,* August 6, 1935.

47. Minutes, KLM board meeting, August 6, 1935: KLM Board Papers, series R-5.

48. Anthony Fokker to the board of KLM, August 22, 1935: KLM Board Papers, series R-5 (Fokker's emphasis and capitalization).

49. Anthony Fokker to KLM's Board of Directors, November 30, 1935: KLM Board Papers, series R-5.

50. Anthony Fokker to the board of KLM, December 16, 1935: KLM Board Papers, series R-5.

51. Van Doorninck to minister of finance, November 27, 1933: Archive of the Ministry of Finance, dossier 1330-4(15). In greater detail, see Dierikx, *Bevlogen jaren,* 64–67. The record of the Van Doorninck Committee is the most important remaining source on the business history of the Nederlandsche Vliegtuigenfabriek. Its hearings and reports are spread over the archives of the Ministry of Finance (section 1330) and that of the Ministry of Waterworks (RLD, 14-III).

52. Minutes, fourth meeting of the Van Doorninck Committee, April 10, 1934: Archive of the Ministry of Finance, dossier 1330-4(16).

53. Stephan and Vattier Kraane to the ministers for colonial affairs and defense, May 19, 1934: Archive of the Ministry of Finance, dossier 1330-6(26).

54. Van Doorninck Report, June 21, 1935: Archive of the Ministry of Waterworks, RLD, 14-III.

55. Anthony Fokker to Van Doorninck, August 16, 1934: Archive of the Ministry of Finance, dossier 1330-6(26).

56. Minutes, ninth meeting of the Van Doorninck Committee, November 8, 1934: Archive of the Ministry of Finance, dossier 1330-4(16).

57. Minutes, eleventh meeting of the Van Doorninck Committee, December 12, 1934: Archive of the Ministry of Finance, dossier 1330-4(16).

58. Balance sheet of the Nederlandsche Vliegtuigenfabriek, drawn up by Tekelenburg & Van Til Accountancy, March 8, 1935: Archive of the Ministry of Finance, dossier 1330-6(25).

59. Minutes of a special meeting of the Van Doorninck Committee with Tony Fokker, February 28, 1935: Archive of the Ministry of Finance, dossier 1330-4(16).

60. Minutes, sixteenth meeting Van Doorninck Committee, May 21, 1935: Archive of the Ministry of Finance, dossier 1330-4(16).

61. Stephan to Hegener, January 5, 1956: Aeronautical Collection, Thijs Postma, Hegener Papers, correspondence.

62. Minutes, twenty-second meeting of the Van Doorninck Committee, January 27, 1936: Archive of the Ministry of Finance, dossier 1330-4(16).

63. Final report of the Van Doorninck Committee, with a concept agreement between the state and Fokker (appendix E) and a concept employment contract with Anthony Fokker personally (appendix D), February 4, 1936: Archive of the Ministry of Finance, dossier 1330-2(6).

64. Chief Treasurer R. A. Ries to minister of finance, February 12, 1936: Archive of the Ministry of Finance, dossier 1330-2(6).

65. Minister of defense to minister of finance, March 12, 1936: Archive of the Ministry of Finance, dossier 1330-2(6).

66. Recollections of Anthony Fokker by Wim van Nijenhoff, March 1, 1940: Aeronautical Collection, Thijs Postma, Hegener Papers, correspondence. See also *Biografisch Woordenboek,* vol. 1, 188.

## CHAPTER 9. ROCKING THE BOAT

1. Rae, *Climb to Greatness,* 65–66.

2. Robert E. Gross to Courtlandt S. Gross, January 23, 1934: U.S. Library of Congress, Manuscript Division, Robert E. Gross Papers, box 4.

3. The acquisition of the rights to the L-10A Electra for 26,372.89 guilders appeared in the books of the Dutch company next to those of the DC-2. Audit by Tekelenburg & Van Til Accountancy, March 7, 1935: Archive of the Ministry of Finance, dossier 1330-6(25).

4. Testimony of Anthony Fokker before the U.S. Senate Special Committee Investigating the Munitions Industry (chaired by Sen. Gerald P. Nye) on September 12, 1935: U.S. National Archives, Record Group 46, Executive file, box 156. Fokker's testimony was later communicated to the press by Senator Nye in October 1936 to put an end to rumors about Elliott Roosevelt's involvement in trade with the Soviet Union. *New York Times,* October 7, 1936.

5. Biddle, *Barons of the Sky,* 203. Biddle quotes a letter that Tony Fokker wrote to Donald Douglas on his European sales trip.

6. Robert E. Gross to Courtlandt S. Gross, May 3, 1934: U.S. Library of Congress, Manuscript Division, Robert E. Gross Papers, box 4.

7. Interview with Anthony Fokker (undated newspaper cutting from the *Handelsblad,* 1934): Algemeen Rijks Archief, 2.21.081, Collectie J. A. van Hamel, no. 223.

8. The term is from Alan Bullock, *Hitler and Stalin: Parallel Lives* (London: Harper Collins, 1991), 285ff.

9. Ibid., 326.

10. Fokker to J. A. van Hamel, September 25, 1934: Algemeen Rijks Archief, 2.21.081, Collectie J. A. van Hamel, no. 222; interview with Anthony Fokker (undated newspaper cutting from the *Handelsblad,* 1934): Algemeen Rijks Archief, 2.21.081, Collectie J. A. van Hamel, no. 223. See also Knapen, *De lange weg naar Moskou,* 164ff. Until 1942 The Hague's fiercely Christian government (exiled after May 1940) refused to officially recognize the USSR because of Moscow's atheist stance.

11. Interview with Anthony Fokker (undated newspaper cutting from the *Handelsblad,* 1934): Algemeen Rijks Archief, 2.21.081, Collectie J. A. van Hamel, no. 223.

12. Testimony of Anthony Fokker before the U.S. Senate Special Committee Investigating the Munitions Industry (chaired by Sen. Gerald P. Nye) on September 12, 1935: U.S. National Archives, Record Group 46, Executive file, box 156.

13. Robert E. Gross to Randolph C. Walker, July 30, 1934: U.S. Library of Congress, Manuscript Division, Robert E. Gross Papers, box 4. Dates from stamps in Fokker's Dutch passport (issued July 28, 1932): Aviodome Archive, Passports Anthony Fokker dossier.

14. Nevil Shute Norway, *Slide Rule: The Autobiography of an Engineer* (London: Heinemann, 1954), chapters 7–9. Airspeed had by then moved to Portsmouth.

15. Norway, *Slide Rule,* 219–21; minutes, thirteenth meeting of the Van Doorninck Committee, February 5, 1935: Archive of the Ministry of Finance, dossier 1330-4(16); report of the Van Doorninck Committee, June 21, 1935: Archive of the Ministry of Waterworks, RLD, 14-III.

16. Minutes, special meeting of the Van Doorninck Committee with Tony Fokker, Vattier Kraane, and F. Elekind, February 28, 1935: Archive of the Ministry of Finance, dossier 1330-4(16).

17. Norway, *Slide Rule,* 229.

18. Summary of correspondence on four-engined aircraft for KLM (compiled by Fokker), February 24, 1939: Archive of the Ministry of Waterworks, RLD, 14-V.

19. Van Tijen (Fokker's new deputy director) to minister of waterworks, January 22, 1936: Archive of the Ministry of Waterworks, RLD, 14-III. See also Lambert Slotemaker (KLM's general secretary) to the board of KLM, September 6, 1935; and the report of Plesman to the board of KLM, November 11, 1935: KLM Board Papers, series R-5. The four-engined project Douglas was working

on eventually became known as the DC-4E. It proved to be overweight and unsuccessful as a result of the enormous technological step Douglas wanted to make with the DC-4E, and it was abandoned in favor of a down-scaled, unpressurized variant, for which Douglas maintained the same type designation: DC-4 (or C-54 Skymaster, in its military guise).

20. Summaries of correspondence on two-engined and four-engined aircraft for KLM (compiled by Fokker), February 24, 1939: Archive of the Ministry of Waterworks, RLD, 14-V.

21. In June and July 1936, the Netherlands Civil Aeronautics Service (Rijks Luchtvaart Dienst) prepared a number of memos for the minister of waterworks spelling out the inability of the Fokker company to license-manufacture the DC-3: Archive of the Ministry for Waterworks, RLD, 14-III. KLM received a total of twenty-three DC-3's from Fokker. Report, KLM Technical Department, May 1, 1939: KLM Board Papers, series R-5.

22. Dierikx, *Bevlogen jaren,* 100–1.

23. A. H. G. Fokker, memorandum on future demands on airport facilities in relation to the plans for the expansion of Schiphol airport (with annexes), May 5, 1936: Amsterdam Municipal Archive, Collection 5183, Handels-Inrichtingen, 1936-300.

24. *New York Times,* July 3, 1930.

25. Ibid., June 20 and July 20, 1931.

26. Carter Tiffany, "A Personal Memory of Tony Fokker," undated (ca. 1940): Aeronautical Collection, Thijs Postma, Hegener Papers, correspondence; Smedley and Noble, "Profiles"; news items on Anthony Fokker in the *New York Times,* September 4, 1935, and January 14, 1938; photographs of one of Tony Fokker's fishing trips, *Het Vliegveld* 24, no. 3 (1940): 52–53.

27. *New York Times,* September 4, 1935.

28. Carter Tiffany, "A Personal Memory of Tony Fokker," undated (ca. 1940): Aeronautical Collection, Thijs Postma, Hegener Papers, correspondence; Wim van Neijenhoff, personal recollections of Anthony Fokker, March 1, 1940: Aeronautical Collection, Thijs Postma, Hegener Papers, correspondence. See also *New York Times,* September 5, 1935. Fokker's testimony was typed up and signed by him on September 18, 1935: U.S. National Archives, Record Group 46, Executive file, box 156.

29. *New York Times,* October 7, 8, 9, 10 and 11, 1936.

30. *Los Angeles Times,* October 7, 1936; *Los Angeles Illustrated Daily News,* October 8, 1936.

31. In June 1936 Fokker had his personal secretary in Amsterdam, Cis van Rhijn, write to Mock telling him not to send so many copies of correspondence because a lot of the matters were "not important to him." Correspondence between Richard Mock and Cis van Rhijn, June 1936: Fokker Archive, A. H. G. Fokker, box 2.

32. Telegrams between Mock and Fokker, June–August 1936: Fokker Archive, A. H. G. Fokker, box 2.

33. Telegram, Mock to Fokker, July 17, 1936; and Mock to Tiffany, July 31, 1936: Fokker Archive, A. H. G. Fokker, box 2.

34. Mock to Fokker at the Hotel U.S. Grant in San Diego, September 12, 1937: Fokker Archive, A. H. G. Fokker, box 2.

35. Correspondence between Fokker and Mock, September–October 1937: Fokker Archive, A. H. G. Fokker, box 2.

36. Mock to the El Tovar Hotel, September 29, 1936: Fokker Archive, A. H. G. Fokker, box 2.

37. Richard Mock to Miss Geneva Carr at the Los Angeles Eye & Ear Hospital, Nov. 4, 1935: Fokker Archive, A. H. G. Fokker, box 2.

38. Mock to Fokker, Sep 20, 1937: Fokker Archive, A. H. G. Fokker, box 2.

39. Carter Tiffany, "A Personal Memory of Tony Fokker," undated (ca. 1940): Aeronautical Collection, Thijs Postma, Hegener Papers, correspondence. See also *New York Times,* January 14, 1938; and E. Franquinet, *Fokker: Een leven voor de luchtvaart* (Eindhoven: De Pelgrim, 1946), 331–32.

40. Mock to Fokker at the Consolidated Shipbuilding Corp. in New York, April 25, 1938: Fokker Archive, A. H. G. Fokker, box 2.

41. Carter Tiffany, "A Personal Memory of Tony Fokker," undated (ca. 1940): Aeronautical Collection, Thijs Postma, Hegener Papers, correspondence.

42. Fokker's speech was quoted completely in the *New York Times,* June 21, 1938.

43. Ibid.

44. Carter Tiffany, "A Personal Memory of Tony Fokker," undated (ca. 1940): Aeronautical Collection, Thijs Postma, Hegener Papers, correspondence. See also Donald W. Douglas, "Remembering Tony Fokker," *Het Vliegveld* 25, no. 1 (1941): 14; *New York Times,* December 24, 1939.

45. Bartlett Gould, "The Last Days of Anthony H. G. Fokker," *World War I Aero* 21, no. 89 (1982): 26–27 (based on the personal papers of Starling Burgess).

46. Küpfer typescript, 139 (based on interviews with J. E. van Tijen).

47. Joh. de Vries, ed., *Herinneringen en dagboek van Ernst Heldring,* vol. 2 (Groningen: Wolters Noordhof, 1970), 114 (diary entry of April 5, 1935).

48. Interview with Tony Fokker in the *New York Times,* March 19, 1938. On the widespread notion that "the bomber will always get through," see Uri Bialer, *The Shadow of the Bomber: The Fear of Air Attack and British Politics, 1932–1939* (London: Royal Historical Society, 1980).

49. Testimony of Carter Tiffany before the Supreme Court of Rockland County, January 29, 1943: Fokker Archive, V/Alg. Secr., no. 13. In the months leading up to the German invasion of Holland in May 1940, the management of the Nederlandsche Vliegtuigenfabriek would add a further $1.5 million to its U.S. cash reserves on the advice of Tiffany, bringing the company's gross assets entrusted to safekeeping in American banks on the eve of the war to 36 percent.

50. *Annual Report NV Nederlandsche Vliegtuigenfabriek Fokker in 1937,* 13: Fokker Archive. With an eye to future expansion, the total possible emissive stock capital was increased as well from 1,500,000 to 10,050,000 guilders.

51. The 1937 figure is from a Fokker advertisement in *Het Vliegveld* 21, no. 8 (1937): v; the 1938 figure is from Joh. de Vries, *Met Amsterdam als brandpunt: Hondervijftig jaar Kamer van Koophandel en Fabrieken, 1811–1961* (Amster-

dam: Ellerman Harms, 1961), 84. In 1937 the factory signed contracts with the Netherlands Ministry of Defense for accelerated delivery of eighty-eight first-line military aircraft, sixteen T.V's, thirty-six D.XXI's, and thirty-six G.1's. Memorandum, minister of defense to the cabinet, January 5, 1938: Archive of the Netherlands Ministry of Finance, dossier 1330-5(18).

52. "De luchtvaart in juni en juli," *Het Vliegveld* 21, no. 8 (1937): 268.

53. Balance and loss statements, *Annual Reports NV Nederlandsche Vliegtuigenfabriek Fokker over 1937, 1938, and 1939:* Fokker Archive.

54. Fokker to Colijn, February 11, 1939: Archive of the Netherlands Ministry of Waterworks, RLD, 14-V (Fokker's emphasis).

55. H. Ch. E. van Ede van der Pals to minister of waterworks, February 16, 1939: Archive of the Netherlands Ministry for Transport and Waterworks, RLD, 14-V.

56. Minister of waterworks to Plesman, March 4, 1939: Archive of the Netherlands Ministry for Transport and Waterworks, RLD, 14- V.

57. Fokker to Plesman, March 9, 1939: KLM Board Papers, series A-9/A-950, R-506.2; protocol of the meeting on March 13, 1939, between KLM and Fokker in the Royal Dutch/Shell office: Archive of the Netherlands Ministry of Transport and Waterworks, RLD, 14-V.

58. *De Telegraaf,* March 15, 1939; apologetic letter of Anthony Fokker to G. W. J. Bruins (a prominent board member of KLM), March 15, 1939: KLM Board Papers, series A-9/A-950, R-506.2.

59. Description of the F.180 project by P. de Winter of the Netherlands Aeronautical Service, February 28, 1940: Archive of the Netherlands Ministry of Finance, dossier 1330-9(42).

60. Contract between the government of the Netherlands and the Nederlandsche Vliegtuigenfabriek Fokker, January 22–23, 1940; contract between the Nederlandsche Vliegtuigenfabriek Fokker and KLM, January 25/26, 1940: Archive of the Netherlands Ministry of Transport and Waterworks, RLD, no. 434.

61. Dierikx, *Bevlogen jaren,* 104–6.

62. Anna Fokker-Diemont to Christine Döppler, January 27, 1940: Aeronautical Collection, Tom Heitzman, Döppler album. See also Gould, "Last Days."

63. *New York Times,* December 22 and 24, 1939.

64. Anna Fokker-Diemont to Christine Döppler, January 27, 1940: Aeronautical Collection, Tom Heitzman, Döppler album.

65. Official photo album of the interring of Fokker's ashes compiled by the NV Nederlandsche Vliegtuigenfabriek Fokker, undated: Aviodome Archive. Jojo de Leuw later returned to the United States as the spouse of Frank Dulles, Carter Tiffany's assistant in the liquidation of Fokker's estate.

# ARCHIVAL SOURCES

## THE HAGUE, NETHERLANDS

**Algemeen Rijks Archief (Netherlands National Archive), 2e Afdeling**
Archive of the Ministry of Finance, 1919–40, papers of the Van Doorninck
  Committee (1330-7)
Archive of the Ministry of Waterworks: Rijks Luchtvaart Dienst, 1920–40
Archive of the Staatscommissie inzake Luchtvaart, 1919–30
Archive of the Commissie van Advies van de Rijks Studiedienst voor Luchtvaart
Collection 2.21.081, J. A. van Hamel
Collection 2.21.043, Cremer Family Papers

**Archive of the Ministerie van Buitenlandse Zaken (Netherlands Foreign Office)**
Netherlands mission in Peking, series V, 1931–41, box P.IV
Directie Economische Zaken (Department of Economic Affairs), DEZ III-336,
  DEZ 70b *Romania*, DEZ 87 *Rusland*, DEZ 151 *Luchtvaart; bepalingen enz.*,
  DEZ 151-44, DEZ 151-45, DEZ 151-92

**Centraal Archievendepot Ministerie van Defensie (Netherlands
  Defense Ministry)**
General Headquarters, dossiers GG.151b
Commando Luchtverdediging (Air Defense Command), CLV, no. 39
Ministry of War, series 2774, Afd. IV, dossier 57

**Royal Netherlands Air Force, Air Historical Branch**
Diary of Col. Hendrik Walaardt Sacré, commander of the Air Corps

**Archive of the Board of Directors of the Royal Netherlands Post and Telecom
  (PTT)**
Verbalenarchief Centrale Directie P&T, 1918–21

**Wm. Teuben "Teubarium," Aeronautical Collection (Voorburg)**
Various personal belongings of Anthony Fokker
Miscellaneous papers of the Comité Vliegtocht Nederland-Indië

## AMSTERDAM AND AMSTERDAM AIRPORT, SCHIPHOL, NETHERLANDS

**Fokker Aircaft BV**
Static archive (Schiphol-Verhoef): series V/Alg. Secr. 67 (six boxes), series KW,
    series VO/PT, typescript commemorative book by C. C. Küpfer on forty years
    of Fokker factory
Fokker factory (Schiphol): annual reports, 1937–39 and 1940–46, dossier
    F.XXII/F.XXXVI
Main office (Amsterdam-Zuidoost): Fokker Historical Collection, A. H. G.
    Fokker and M. Beeling dossiers

**Aviodome Museum Collection**
Anthony Fokker dossiers, personal correspondence with his family
Fokker-Germany dossier
Netherlands passports of Anthony Fokker
Miscellaneous data on Fokker Aircraft
ELTA photo album
Official photo album, interring of Fokker's ashes

**Koninklijke Luchtvaart Maatstchappij (KLM Royal Dutch Airlines),
    Amstelveen**
KLM Board Papers, 1919–40, series R-5, R-501, R-506.2,
    A-9, A-950

**Amsterdam Airport, Schiphol**
Documentation Department, annual reports, 1926–40
Miscellaneous correspondence

## HAARLEM, NETHERLANDS

**Municipal Archive**
New Archive, no. 126-1 (Netherlands citizenship question)

**Haarlem Town Hall**
Civil Registration Department, Marriage Certificates, 1919–94

## HOOFDDORP, NETHERLANDS

**Aeronautical Collection, Thijs Postma**
Hegener Papers
Miscellaneous files on Anthony Fokker and the Fokker Aircraft Factory

## UTRECHT, NETHERLANDS

**Archive of the Nederlandse Spoorwegen (Netherlands Railways)**
Board of Control (Raad van Commissarissen), 1918–19

## MUNICH, GERMANY

**Deutsches Museum**
Junkers Archives: Junkers Flugzeugbau im 1· Weltkrieg, 1916–17, JA 0201/T9,
   JA 0201/T11, JA 0201/T12

## LONDON, ENGLAND

**Public Records Office (Kew)**
Foreign Office papers (FO.371), Inter-Allied Commissions of Control,
   Germany
War Trade Intelligence Department records (TS.14), Contraband Committee

## WASHINGTON, D.C., UNITED STATES

**U.S. National Archives**
U.S Senate, Record Group 46, executive file, Nye Committee
State Department, Record Group 84, Netherlands Post File, general
   correspondence
Federal Bureau of Investigation (FBI), Record Group 65, Microfilms M1085,
   roll 00779

**Library of Congress, Manuscript Division**
Robert E. Gross Papers
William B. Mitchell Papers
Edward V. Rickenbacker Papers

**Smithsonian Institution, National Air and Space Museum (NASM)**
Garber Facility, A. Francis Arcier Collection
Aeronautics Department: files Fokker, Anthony H.G. (CF 372000); Platz,
   Reinhold (CP 260000); Noorduyn, Robert B. C. (CN 083000); Rockne
   Crash file
NASM film archive, Fox Movietone reel on Anthony Fokker

## PRINCETON, NEW JERSEY, UNITED STATES

**Aeronautical Collection, Peter M. Grosz**
Fokker files: miscellaneous correspondence, statistics, technical data on World
   War I
Weyl-Platz correspondence

## DEANSBORO, NEW YORK, UNITED STATES

**Tom Heitzman "Stuffinder" Collection**
Photo albums of Christine Döppler, containing letters from Tony Fokker and
   Anna Fokker-Diemont to Döppler
Miscellaneous information on the Atlantic Aircraft Corporation and the Fokker
   Aircraft Corporation

```
╔══════════════════════════════╗
║      BIBLIOGRAPHY            ║
╚══════════════════════════════╝
```

## PERIODICALS

*Aeronautica: Official Publication for Historical Associates of the Institute of Aeronautical Sciences* (selected issues)
*Aeronautics* (selected issues)
*Air Enthusiast* (selected issues)
*Avia* (selected issues)
*Aviation* (selected issues)
*Avia/Vliegwereld* 1 (1952)–10 (1961)
*Fokker Bulletin* 1 (1924)–16 (1940)
*Het Handelsblad* (selected issues)
*Het Vliegveld* 1 (1917)–25 (1941)
*Journal of the American Aviation Historical Society* (selected issues)
*De Maasbode* (selected issues)
*Mededeelingen over het luchtverkeer in en naar Nederlandsch-Indië, uitgegeven door de vereeniging Comité Vliegtocht Nederland-Indië* 1–9 (1924–31)
*New Yorker,* 1929–31
*New York Times,* 1920–40
*Technology and Culture* (selected issues)
*Vliegwereld* 1 (1935/36)–6 (1940/41)
*Windsock International* (selected issues)
*World War I Aero* (selected issues)

## BOOKS

*Aircraft Year Books*. New York: Aeronautical Chamber of Commerce of America, 1920–1940.

Alting, Peter. *Van Spin tot Fokker 100: Fokkers civiele vliegtuigen*. Sassenheim: Rebo Producties, 1988.

———. *Fokkers in uniform: Driekwart eeuw militaire Fokkervliegtuigen*. Sassenheim: Rebo Producties, 1988.

American Social History Project. *Who Built America? Working People and the Nation's Economy, Politics, Culture and Society*. Vol. 2, *From the Gilded Age to the Present*. New York: Pantheon Books, 1992.

Baker, Simonds & Co. *The Aviation Industry*. Detroit: Baker, Simonds, 1929.

Balchen, Bernt. *Come North with Me*. New York: Dutton, 1958.

Bergius, C. C. *Die Strasse der Piloten: Die abenteuerliche Geschichte der Luftfahrt*. München: Wilhelm Heyne Verlag, 1971.

Bialer, Uri. *The Shadow of the Bomber: The Fear of Air Attack and British Politics, 1932–1939*. London: Royal Historical Society, 1980.

Biddle, Wayne. *Barons of the Sky: From Early Flight to Strategic Warfare: The Story of the American Aerospace Industry*. New York: Simon & Schuster, 1991.

Bowers, Peter M. *The Fokkers of World War I*. New York: Hobby Helpers Library, 1960.

Boyd, Alexander. *The Soviet Air Force since 1918*. London: Macdonald and Jane's, 1977.

Brooks, Peter W. *The Modern Airliner: Its Origins and Development*. London: Putnam, 1961.

Bullock, Alan. *Hitler and Stalin: Parallel Lives*. London: Harper Collins, 1991.

Byrd, Richard E. *Skyward: Man's Mastery of the Air as Shown by the Brilliant Flights of America's Leading Air Explorer: His Life, His Thrilling Adventures, His North Pole and Transatlantic Flights, together with His Plans for Conquering the Antarctic by Air*. New York: Halcyon House, 1928.

Carreras, Albert, Andrea Giuntini, and Michèle Merger, eds. *European Networks, Nineteenth–Twentieth Centuries: New Approaches to the Formation of a Transnational Transport and Communications System*. Proceedings Eleventh International Economic History Congress, vol. B.8. Milan: Università Bocconi, 1994.

Cashman, Sean. *America in the Twenties and Thirties: The Olympian Age of Franklin Delano Roosevelt*. New York: New York University Press, 1989.

Charité, J., ed. *Biografisch Woordenboek van Nederland*. Vol. 1. The Hague: Nijhoff, 1979.

Corn, Joseph J. *The Winged Gospel: America's Romance with Aviation, 1900–1950*. New York: Oxford University Press, 1983.

Daley, Robert. *An American Saga: Juan Trippe and His Pan Am Empire*. New York: Random House, 1980.

Davies, R. E. G. *A History of the World's Airlines*. London: Oxford University Press, 1964.

————. *Airlines of the United States since 1914*. London: Putnam, 1972.

Davis, Burke. *The Billy Mitchell Affair*. New York: Random House, 1967.

de Jonge, Alex. *The Weimar Republic: Prelude to Hitler*. London: Paddington Press, 1978.

Dierikx, Marc. *Bevlogen jaren: Nederlandse burgerluchtvaart tussen de wereldoorlogen*. Houten: Unieboek, 1986.

————. *Begrensde horizonten: De Nederlandse burgerluchtvaartpolitiek in het Interbellum*. Zwolle: Tjeenk Willink, 1988.

van Dulm, J. F., and F. C. van Oosten. *50 jaar Marine Luchtvaart Dienst, 1917–1967*. The Hague: Koninklijke Marine, 1967.

Fahey, James C. *U.S. Army Aircraft (Haevier Than Air), 1908–1946*. New York: Ships and Aircraft, 1978.

*Financial Handbook of the American Aviation Industry—July 1929*. New York: Commercial National Bank and Trust Company, 1929.

Fischer von Poturzyn, A. D. *Junkers und die Weltluftfahrt: Ein Beitrag zur Entstehungsgeschichte Deutscher Luftgeltung, 1909–1933*. München: Richard Pflaum Verlag, 1933.

Fischer, Fritz. *Griff nach der Weltmacht: Die Kriegszielpolitik des kaiserlichen Deutschland, 1914–18*. Düsseldorf: Droste Verlag, 1961.

Fokker, Anthony H. G., and Bruce Gould. *Flying Dutchman: The Life of Anthony Fokker*. New York: Henry Holt, 1931.

————. *De Vliegende Hollander*. Amsterdam: Van Holkema en Warendorf, 1931.

————. *Souvenirs d'un homme volant: La vie d'Anthony Fokker*. Paris: Calman-Lévy, 1932.

————. *Der fliegende Holländer: Das Leben des Fliegers und Flugzeugkonstrukteurs A. H. G. Fokker*. Zürich: Rascher & Cie. Verlag, 1933.

Forden, Lesley. *The Ford Air Tours, 1925–1931: A Complete Narrative and Pictorial History of the Seven National Air Tour Competitions for the Edsel B. Ford Reliability Trophy*. Alameda, Calif.: Nottingham Press, 1973.

Franquinet, E. *Fokker: Een leven voor de luchtvaart*. Eindhoven: De Pelgrim, 1946.

Freudenthal, Elsbeth E. *The Aviation Business: From Kitty Hawk to Wall Street*. New York: Vanguard Press, 1940.

Freund, G. *Unholy Alliance: Russian-German Relations from the Treaty of Brest-Litovsk to the Treaty of Berlin*. London: Chatto and Windus, 1957.

Fritsche, Peter. *A Nation of Fliers: German Aviation and the Popular Imagination*. Cambridge: Harvard University Press, 1992.

Grey, C. G., ed. *Jane's All the World's Aircraft*. London: Sampson, Low, Marston, 1910/11–1940.

Grosz, Peter M., George Haddow, and Peter Schiemer. *Austro-Hungarian Army Aircraft of World War One*. Mountain View, Calif.: Flying Machine Press, 1994.

Guilonard, Pieter, and Gijs Spit. *De weg door de lucht: Het moderne snelverkeer*. Amsterdam: Scheltens & Giltay, 1927.

Hansen, James R. *Engineer in Charge: A History of the Langley Aeronautical*

*Laboratory, 1917–1958*. Washington, D.C.: National Aeronautics and Space Administration, 1987.

Hegener, Henri. *Fokker: The Man and the Aircraft*. Letchworth: Harleyford Publications, 1961.

Hermans, Frans J., ed. *Venlo's Mozaïek: Hoofdstukken uit zeven eeuwen stadsgeschiedenis*. Maastricht: Limburgs Geschied- en Oudheidkundig Genootschap, 1990.

Holden, Henry M. *The Fabulous Tri-Motors*. Blue Ridge Summit, Penn.: TAB Books, 1992.

Homze, Edward L. *Arming the Luftwaffe: The Reich Air Ministry and the German Aircraft Industry, 1919–1939*. Lincoln: University of Nebraska Press, 1976.

Hooftman, Hugo. *Fokker: Neerlands grootste vliegtuigbouwer*. Alkmaar: Arti, 1959.

Imrie, Alex. *The Fokker Triplane*. London: Arms & Armour Press, 1992.

Ingells, Douglas J. *The McDonnell-Douglas Story*. Fallbrook, Calif.: Aero Publishers, 1979.

Ingold, Felix Ph. *Literatur und Aviatik: Europäische Flugdichtung, 1909–1927*. Basel: Birkhäuser Verlag, 1978.

de Jong, A. P., ed. *Vlucht door de tijd: 75 jaar Nederlandse Luchtmacht*. Houten: Unieboek, 1988.

de Jong, Lou. *Het Koninkrijk der Nederlanden in de Tweede Wereldoorlog*. Vol. 1, *Voorspel*. The Hague: Staatsuitgeverij, 1969.

———. *Het Koninkrijk der Nederlanden in de Tweede Wereldoorlog*. Vol. 11a, *Nederlands-Indië—I, Eerste Helft*. The Hague: Staatsuitgeverij, 1984.

Josephson, Matthew. *Empire of the Air: Juan Trippe and the Struggle for World Airways*. New York: Harcourt, Brace, 1944.

von Kármán, Theodore. *The Wind and Beyond: Pioneer in Aviation and Pathfinder in Space*. Boston: Little, Brown, 1967.

Kingsford-Smith, Charles E., and Charles T. P. Ulm. *The Flight of the Southern Cross*. New York: McBride, 1929.

Knapen, Ben. *De lange weg naar Moskou: Nederlandse relaties tot de Sovjet Unie, 1917–1942*. Amsterdam: Elsevier, 1985.

Komons, Nick A. *Bonfires to Beacons: Federal Aviation Policy under the Air Commerce Act, 1926–1938*. Washington, D.C.: Smithsonian Institution Press, 1989.

Koppen, George A. *Holland-Indië-Holland, in strom en zonnebrand*. Amsterdam: Telegraaf, 1927.

Kranzhof, Jörg Armin. *Flugzeuge die Geschichte machten: Fokker Dr.I*. Stuttgart: Motorbuch Verlag, 1994.

Leavitt, Judith Walzer, ed. *Women and Health in America: Historical Readings*. Madison: University of Wisconsin Press, 1984.

de Leeuw, René, ed. *Fokker verkeersvliegtuigen: Van de F.I uit 1918 tot en met de Fokker 100 van nu*. Amsterdam: NV Koninklijke Nederlandse Vliegtuigenfabriek Fokker, 1989.

Loening, Grover. *Our Wings Grow Faster*. Garden City, N.Y.: Doubleday, 1935.

Meijer, J. C., ed. *Nederlandsche Staatswetten*. Sneek: Van Druten & Bleeker, 1893.

Meulen, Jacob A. Vander. *The Politics of Aircraft: Building an American Military Industry*. Lawrence: University of Kansas Press, 1991.

Meyer, Henry Cord. *Airshipmen, Businessmen, and Politics, 1890–1940*. Washington, D.C.: Smithsonian Institution Press, 1991.

Mommsen, Hans. *Die verspielte Freiheit: Der Weg der Republik von Weimar in den Untergang, 1918 bis 1933*. Ullstein/Propyläen Verlag: Frankfurt am Main, 1990.

Montague, Richard. *Oceans, Poles, and Airmen: The First Flights over Wide Waters and Desolate Ice*. New York: Random House, 1971.

Morrison, Wilbur H. *Donald W. Douglas: A Heart with Wings*. Ames: Iowa State University Press, 1991.

Morrow, John H., Jr. *German Air Power in World War I*. Lincoln: University of Nebraska Press, 1982.

———. *The Great War in the Air: Military Aviation from 1909 to 1921*. Washington, D.C.: Smithsonian Institution Press, 1993.

Munson, Kenneth. *Historische Vliegtuigen, 1903–1914*. Amsterdam: Moussault, 1970.

———. *Airliners between the Wars, 1919–1939*. London: Blandford Press, 1972.

Newton, Wesley P. *The Perilous Sky: Evolution of United States Aviation Diplomacy toward Latin America, 1919–1931*. Coral Gables, Fla.: University of Miami Press, 1978.

van Neyenhoff, W. M., and C. W. A. Oyens. *Zweefvliegen: Theorie en praktijk*. Amsterdam: Meulenhoff, 1935.

van de Noort, Peter F. A. *Fokkers "Roaring Twenties": De vliegtuigen van de Amerikaanse Fokker-fabrieken*. Sassenheim: Rebo Produkties, 1988.

Norway, Nevil Shute. *Slide Rule: The Autobiography of an Engineer*. London: Heinemann, 1954.

Nowarra, Heinz J. *Die verbotenen Flugzeuge, 1921–1935: Die getarnte Luftwaffe*. Stuttgart: Motorbuch Verlag, 1980.

Parmentier, Koene D. *In drie dagen naar naar Australië: Met de Uiver in de Melbourne Race*. Amsterdam: Scheltens & Giltay, 1935.

*Persoonlijkheden in het Koninkrijk der Nederlanden in Woord en Beeld*. Amsterdam: Van Holkema & Warendorf, 1938.

Plesman, Albert, Jr. *Albert Plesman, mijn vader*. The Hague: Nijgh & Van Ditmar, 1977.

Pollog, Carl H. *Hugo Junkers: Ein Leben als Erfinder und Pionier*. Dresden: Carl Reissner Verlag, 1930.

Postma, Thijs. *Fokker: Aircraft Builders to the World*. London: Jane's Publishing, 1980.

Puffer, Claude E. *Air Transportation*. Philadelphia: Blakiston, 1941.

Rae, John B. *Climb to Greatness: The American Aircraft Industry, 1920–1960*. Cambridge: MIT Press, 1968.

Rickenbacker, Edward V. *Rickenbacker: An Autobiography*. Englewood Cliffs, N.J.: Prentice-Hall, 1967.

Roeder, Jean. *Bombenflugzeuge und Aufklärer: Von der Rumpler-Taube zur DO.23*. Bonn: Bernard & Graefe Verlag, 1990.

Schmitt, Günter. *Hugo Junkers und seine Flugzeuge*. Stuttgart: Motorbuch Verlag, 1986.

Simonson, Gene R., ed. *The History of the American Aircraft Industry: An Anthology*. Cambridge: MIT Press, 1968.

Sloan, Alfred P., Jr. *My Years with General Motors*. Garden City, N.Y.: Doubleday, 1963.

Smirnoff, Iwan. *Smirnoff vertelt*. Amsterdam: Blitz, 1938.

Smith, Henry L. *Airways: The History of Commercial Aviation in the United States*. Reprint, Washington, D.C.: Smithsonian Institution Press, 1991.

Smithsonian Institution National Air and Space Museum. *Milestones of Aviation*. New York: Hugh Lauter Levin, 1989.

Solberg, Carl. *Conquest of the Skies: A History of Commercial Aviation in America*. Boston: Little, Brown, 1979.

Trippe, Betty Stettinius. *The Diary and Letters of Betty Stettinius Trippe, 1925–1968*. New York: privately published, 1982.

Suwelack, Winfried and Walter. *Josef Suwelack und der Traum vom Fliegen: Leben und Tod des Billerbecker Flugpioniers und Weltrekordlers Josef Suwelack, 30. April 1888 bis 13. September 1915*. Warendorf: Karl Darpe Verlag, 1988.

van der Hoop, A. N. J. Thomassen à Thuessinck. *Door de lucht naar Indië*. Amsterdam: Scheltens & Giltay, 1924.

Trimble, William F., ed. *Pioneers and Operators*. Vol. 2 of *From Airships to Airbus: The History of Civil and Commercial Aviation*. Washington, D.C.: Smithsonian Institution Press, 1995.

van Oyen, A. A. Vorsterman. *Stam- en Wapenboek der aanzienlijke Nederlandsche familiën*. Vol. 1. Groningen: Wolters, 1885.

de Vries, G., and B. J. Martens. *Fokker Vliegtuigbewapening*. Amsterdam: De Bataafsche Leeuw, 1994.

de Vries, Joh. *Met Amsterdam als brandpunt: Hondervijftig jaar Kamer van Koophandel en Fabrieken, 1811–1961*. Amsterdam: Ellerman Harms, 1961.

————, ed. *Herinneringen en dagboek van Ernst Heldring (1871–1954)*. Groningen: Wolters Noordhof, 1970.

Weiss, David A. *The Saga of the Tin Goose: The Plane that Revolutionized American Civil Aviation*. New York: Crown Publishers, 1971.

Wennekes, Wim. *De aartsvaders: Grondleggers van het Nederlandse bedrijfsleven*. Amsterdam: Uitgeverij Atlas, 1993.

Weyl, A. R. *Fokker: The Creative Years*. London: Putnam, 1965.

Wilkins, George H. *Flying the Arctic*. New York: Putnam, 1928.

Wohl, Robert. *A Passion for Wings: Aviation and the Western Imagination, 1908–1918*. New Haven: Yale University Press, 1994.

Woltjer, J. J. *Recent verleden: Nederland in de twintigste eeuw*. Amsterdam: Uitgeverij Maarten Muntinga, 1994.

# IND-EX

241